The Fragrance of Paradise

The Fragrance of Paradise

Stephen Hoy

The Fragrance of Paradise

By: Stephen Hoy

Copyright © 2009 by Stephen Hoy
All rights reserved.

This book or parts thereof may not be reproduced in any form, stored in a retrieval system, or transmitted in any form by any means—electronic, mechanical, photocopy, recording, or otherwise—without prior written permission of the publisher, except as provided by United States of America copyright law.

All Bible References from the NIV unless otherwise noted.
HOLY BIBLE, NEW INTERNATIONAL VERSION ®
Copyright © 1973, 1978, 1984 by International Bible Society
Used by permission of Zondervan Publishing House.
All rights reserved.

Published by:
Crossover Publications LLC
870 N. Bierdeman Road
Pearl, Mississippi 39208
www.crossoverpublications.com
Phone: (601) 664-6717, Fax: (601) 664-6818

Library of Congress Control Number: 2009933904

ISBN 978-0-9819657-2-7

Printed in the USA

Cover design and layout: R. Matthew Mooney
Cover concept and artwork: Randall M. Mooney

TABLE OF CONTENTS

INTRODUCTION

CHAPTER ONE
HOMESICK FOR EDEN
-1-

CHAPTER TWO
HISTORICAL EVIDENCE
-7-

CHAPTER THREE
THE FRAGRANT TOUCH OF GOD'S SPIRIT
-13-

CHAPTER FOUR
ADORNED IN THE FRAGRANCE OF THE KNOWLEDGE OF CHRIST
-33-

CHAPTER FIVE
STEPPING STONES THAT LEAD TO THE FRAGRANCE OF GOD'S PRESENCE
-61-

CHAPTER SIX
THE SECRET PLACE
-105-

APPENDIX

ACKNOWLEDGEMENTS

Any thanks must begin with the person who saved my life, my lovely wife Rita. It was her faith, her prayers, and her love that moved God's heart. Joining her voice was that of my parents, my brothers and their families, Rita's family, my BMFIU - Bill Murphy, Steve and Lynn Sieting, Kathy and Keith Trout, and my First Assembly of God family.

Thank you to Mike Ludicke, Alan Grimes, Joe Meadows, David Williams, Chuck Chalk, and David Kibler, for their faith-filled prayers. To Buzz and Linda Middleton, thanks for walking Rita and me through the liver transplant process. The well-wishes and prayers of my Houston County School System and Air Force Band families poured in for months and were a huge source of comfort. Gratitude is also due the doctors, nurses and staff of the Emory Transplant Clinic, and the medical staff on the 9th floor of Emory University Hospital.

I would like to offer a special thank you to the spiritual leaders that have played such a large role in shaping my spiritual development: Ernie Reisinger, Walt Chantry, Dr. John Werner, Jim Eubanks, Dan Powell, Ken Craig, Kenneth Broadus, Richard Black, Clyde Johns, Glynn Grantham, and Mark Merrill.

Lastly, thank you to Malaudi Cassandro for transcribing hours of teaching tapes, and to Dr. Dawn Owens for invaluable editorial assistance.

"Let my well-being be a roar of thanksgiving-filled praise and joyful shouting that can never be silenced."

Introduction

*Close your eyes. Be still and take a deep breath.
Inhale the fragrance of my presence.
Let your restlessness be replaced with the
Stillness of my peace.*

For months and months, I watched myself deteriorate physically. A persistent question repeatedly filled my thoughts, "How sick am I going to get?" One sleepless night, in the midst of a moment of physical and emotional despair, I cried out to God for relief and for his healing touch. Quietly and gently, God's Holy Spirit responded to my desperate cry.

My mind was instantly quieted as if I had come upon an unexpected stop sign. As if searching for some familiar scent, I began to inhale, slowly filling my lungs with air. The memory of some of my favorite fragrances came to mind, including the scent of roses and gardenias. Then I remembered an overwhelming late night drive through a Florida orange grove in full blossom. In a bus full of Air Force Band friends and coworkers, all conversation came to an amazing halt as the incredibly fragrant aroma drifted through the bus's open windows.

What and how does fragrance have anything to do with the comforting presence of God? Despite my inability at that moment to answer these questions, I began to faithfully follow God's admonition to breathe in the stillness of his peace. Hope began to replace my fears and uncertainties. Quiet confidence and trust in my God led me into a place of peace and strength that enabled me to endure what was to come in just a few short months.

In mid January of 2004, I fell into a coma twice in the space of one week's time as a result of near liver failure. The second and more life-threatening coma lasted almost three days. Upon awakening, I received a miraculous gift of life in the form of a liver transplant. During my recovery, I began searching the scriptures and writing so that the powerful message shared with me by God's Holy Spirit might also lead you to a place where you find yourself resting in *the fragrance of paradise.*

> *God Almighty first planted a garden.*
> *And indeed it is the purest of human pleasures.*
> *It is the greatest refreshment to the spirits of man;*
> *without which buildings and palaces are but gross*
> *handy-works: and a man shall ever see that when ages*
> *grow to civility and elegancy, men come to build stately*
> *sooner than to garden finely; as if gardening were*
> *greater perfection.*[1]
>
> Sir Frances Bacon

CHAPTER ONE

HOMESICK FOR EDEN

It has never ended, and in the ongoing history of humankind, it never will end. The overwhelming desire to return to the garden, to paradise, has preoccupied our attention from the moment we were exiled. The evidence can be seen in the literary arts, visual arts, the dramatic arts, and in our music from classical to contemporary. In her song, "Woodstock," Joni Mitchell writes, "We got to get ourselves back to the garden."[2]

Numerous versions and depictions of paradise are present in the enormous variety of artistic, philosophical, and theological endeavors dedicated to the subject. Despite this, several characteristics of "the garden" remain constant.

Each of our senses longs for its own particular satisfaction in our imagined depictions of paradise. The warmth of sunshine, the cool respite of shade, the sound of ocean waves gently splashing, and the touch of a refreshing breeze all bring us pleasure and soothing relaxation. The verdant greens of healthy plants, the vivid blues of the sky and its mirrored reflection in streams and lakes, and the vibrant colors and delicious aromas of flowers and herbs delight us.

Clearly, "the garden" is meaningful to our physical senses! It is also powerfully meaningful to our aesthetic sense, our perception and appreciation of all things beautiful. Additionally, we seem to have an innate understanding that it should be a place of well-being, an emotional oasis of peace and contentment where all the voices that scream for our attention are hushed. We also long for it to be a haven for our rational sense, where our perspective is enlarged and a truthful picture of reality and what is really important is renewed; and a place where frantic activity is replaced with a more aerobic-paced life or even a quiet stillness that leaves room for meditation and reflection.

The energy, time, and resources we spend seeking to get back to the garden reveal the underlying value we attach to this endeavor!

Unfortunately, history reveals that our efforts to recreate paradise fall flat. This is not because our recreations fail to be physically pleasurable or because they fail to provide our souls with the opportunity for rational or emotional refreshing. Rather, the failures result from our ignorance of, or refusal to accept, the fact that when God created mankind and placed Adam and Eve in the garden, Eden was meant to be a place that was also spiritually meaningful. In addition to the beauty and majesty of creation, we were meant to enjoy all the benefits of God's presence—benefits that include the shelter and security of his protection, the provision of our daily bread, the nurture of his love, and an enlightened knowledge of the truth that is found in his presence.

Despite this, the spiritual component is almost universally left out. Or, in order to placate the emptiness that looms large in lives deprived of God's presence, we create miserably pathetic, counterfeit spiritual versions of paradise that are distorted caricatures of the true nature of the garden. As a result, utopian recreations of heaven-on-earth universally fail.

Why? We were kicked out long ago, and our opportunity to dwell there, to establish residency, was revoked. Subsequently, our memory has become vague, and the ruler of this present world has sold us a lie. From the moment we as humans chose to eat of the tree of the knowledge of good and evil, we were banished from the garden and, more significantly, separated from the presence of God. Rather than dwelling in the cool of the garden, contentedly tending trees and plants that are pleasing to the eye and good for the food they provide, rather than peacefully living amongst all the creatures God created, rather than enjoying fellowship with our creator, we find ourselves on the outside looking in. Separated from his presence, we struggle with disease and death. We are both at war with, and fearful of, his creation, and our daily lives are overcome with thorns and thistles, weeds, and painful toil.

It is not only people who have never known God who find themselves in this position. I found myself, a child of God, both emotionally unsettled and struggling to find rest. I was having more and more difficulty dealing with my circumstances, wrestling with sickness, uncertainty, and the uncomfortable squeeze of despair. More and more, I felt like I was losing the battle.

During the summer of 1998, I had been diagnosed with the viral disease Hepatitis C. Through a series of questions and answers, my physicians attributed the infection to a blood transfusion I had received in 1977. Despite undergoing interferon therapy, the disease had progressed to the point of chronic cirrhosis by 2003. As I went through the medical evaluation to become eligible for a liver transplant, I saw some very ill people, many much sicker than I. While visiting the liver transplant clinic at Emory Hospital in Atlanta, Georgia, the realization dawned on me that I was probably looking at my future. In addition to the physical changes taking place as a result of the liver disease, toxic levels of ammonia were robbing my ability to concentrate, to read, and to write or work at the computer. I often misspelled words, wrote them in the wrong order, or left out some words altogether when attempting to write in my journal. There was an ongoing conspiracy in the spirit realm to make me believe God had abandoned me.

Despite these miserable circumstances, I knew from the great spiritual heritage that God had blessed me with while growing up, and through my personal relationship with Jesus Christ, that God was my refuge and strength; he was my peace, my protector, my deliverer, my savior and my healer. Only if I turned my face to him, seeking his presence, would I find what I so desperately needed. When God's still, soft voice directed me to inhale the fragrance of his presence, my journey back to the garden began anew.

What is it that prompts the struggling believer or the person who has never accepted Jesus Christ as LORD and Savior to begin the journey? The prophet Jeremiah tells us that he has placed in all humanity an awareness of, and an elementary knowledge of the truth concerning his existence and character.

I will put my law in their minds and write it in their hearts. No longer will a man teach his neighbor or a man his brother, saying, 'Know the LORD,' because they will know me from the least of them to the greatest, declares the LORD. (Jer. 31:33-34)

This rudimentary awareness that is implanted in our hearts and minds ultimately provokes all mankind to long for the garden. The fact that we live in a harsh world, often filled with conflict, sorrow, loneliness, and sickness, is in a larger sense evidence of our spiritual need for God. Why? It is because only in his presence are ultimately found true peace, rest, joy, companionship, fulfillment, and well-being. Somehow, we have an intuitive understanding that such a place is separate from the world we labor in, and it must be sought. Its desirability powerfully draws us toward a reuniting with God, fulfilling his intention that we dwell with him in the garden, enjoying a reverent fellowship with him, and becoming a worshipping reflection of all of who and what he is.

For those who stubbornly refuse to see or accept the spiritual evidence as well as for those who have accepted the conviction of God—but, in trying times have allowed their eyes to only see their circumstances, God has placed exquisite reminders right in front of us that engagingly draw us toward this longed for goal of returning to the garden. Imagine the brilliance of the rising sun emerging from an ocean-filled horizon, the shadows of majestic clouds moving across a fertile green valley, the grandeur of a forested mountaintop, the glory of sunset-rich colors displayed at the close of day, and the mystery of an infinitely star-lit expanse of midnight sky. All these scenes sing of God's handiwork and present dramatic evidence of his desire to draw us back to a place of nearness to him, the one who created us and knows us by name. Although this journey will only be perfectly realized when we enter a new heaven and a new earth, his placement of a sure knowledge of himself within us is designed to energize a longing in all of humankind to begin our return to the garden even while here on this earth.

The value of a botanic garden was that it conveyed a direct knowledge of God. Since each plant was a created thing, and God had revealed a part of himself in each thing created, a complete collection of each thing created must reveal God completely.[3]

John Prest

CHAPTER TWO

HISTORICAL EVIDENCE

For eleven years of my life, I worked in a family business as a landscape designer. During that time, I developed a lasting interest in the incredible diversity of the flowering trees, shrubs, perennials, and annuals that I worked around. A working knowledge of their sequence of bloom became the predominant feature of my landscape designs. As I also became increasingly familiar with the incredible fragrance of many of these plants, incorporating fragrance became an additional trademark. The first outdoor physical activity I engaged in after recovering from the transplant was restoring my rose garden.

I was already familiar with the important role gardens, both ornamental and agricultural, play in our lives. Historical evidence witnesses man's intuitive understanding that gardens hold some clue to finding peaceful rest, relief from distress, and well-being. Hieroglyphs and paintings found on architectural remains point to long-standing human efforts to recreate Eden.

The peoples of ancient civilizations including Egypt, Sumeria, Assyria, Greece, and Rome established large-scale public and smaller private gardens, the most famous being King Nebuchadnezzar's Hanging Gardens of Babylon, one of the Seven Great Wonders of the World. The word paradise comes from the Persian word *pairidaeza*,[4] meaning "enclosed garden," and was used to describe the royal gardens of Persian King Cyrus.

Spiritually symbolic features were almost always incorporated along with the physical. Efforts to create scenes characterized by peaceful serenity and deliberate simplicity, using evergreen and flowering plant materials scaled down in size, are seen in far eastern gardens. Islamic culture produced gardens containing life-sustaining fruit- bearing trees, shrubs and vines, moving water, rational symmetry, and walls to keep thieves and predators out and to create shelter from the dry and scorching heat of the desert.[5]

Before Christian culture began to have any prevailing influence over Western European culture, monasteries and convents became the birthplace for cloistered physic gardens containing herbs and plants thought to have medicinal properties. These walled gardens allowed their caretakers to cultivate plants from other geographic locations in climates that might not be hospitable to them normally.

Some measure of the geographic search and discovery that was so prevalent in the 15th century, not only sought to prove new scientific theories about the shape of earth and satisfy the quest for riches waiting to be found in other lands, also held out the hope of possibly discovering the Garden of Eden itself. The failure to discover Eden led to the idea of recreating large-scale garden versions of paradise across the European continent and later came to be known as botanic gardens.

Evidence of deliberate spiritual influence in the creation of these gardens is found in several quotes dating back to that era:

> *The re-creation of paradise involves making gardens as near as we can contrive them to resemble the Garden of Eden.*[6]

And,

> *the creation and appreciation of gardens, conveyed a direct knowledge of God. Since each plant was a created thing, and God had revealed himself in each thing created, a complete collection of each thing created must reveal God completely.*[7]

Early large-scale gardens contained a variety of plants, but were predominantly comprised of fruit- producing trees, bushes, and vines planted at the center of the garden or at the corners. Evergreens served either to screen the garden or to create hedge-like borders, while herbs and spices were useful for flavoring, cooking, and making medicines. A water feature was usually present in the form of a fountain or pool. Gradually gardens were also filled with a profusion of ornamental flowering plants. Vast parks planted with grass were also incorporated into the scheme, especially in the largest and most formally designed gardens. This diversity served two purposes: the first was to provide a perpetual succession of flower and/or fruit production, and the second was to mimic a spiritual season of spring-like rebirth and a perpetual season of harvest.

These descriptions mirror biblical references in which God reveals himself to, or is found by, humankind. The prophet Jeremiah creates a vivid picture with these words:

> *I remember the devotion of your youth, how as a bride you loved me and followed me through the desert, through a land not sown...I brought you into a fertile land to eat its fruit and rich produce.*
> (Jer. 2:2, 7)

Israel's first glimpse into this Promised Land in Chapter 14 of Numbers reveals their discovery of a rich harvest, so impressive that it required two men to carry out a cluster of grapes found on one branch. The spies reported that, as God had promised, it indeed "flowed with milk and honey," a reference to a common diet relied upon by nomadic peoples consisting of goat's milk and a "honey-like" syrup pressed from fresh dates.

David writes and sings of God's faithful provision in the richness of green pastures and the refreshing effect of gently flowing streams of water in Psalm 23.

Solomon in his Song of Songs repeatedly alludes to the delights found and produced in the garden including: the pleasant fragrance of perfumes, herbs, and spices (1:3); the adornment of clusters of flowers (1:14); the nourishment of a bountiful harvest of fruit (2:3, 5; 5:1); and the life giving stream of water flowing from a garden fountain (4:15). His beloved compares the arrival of the bridegroom to the arrival of spring, the blossoming of flowers, the initial setting of fruit on the trees, the singing of doves, and the delicious scent of blooming vines filling the air. He can be found in his garden in beds of spices, browsing among and gathering the flowers that grow there (6:2).

Isaiah prophecies of a future time in which the desert will come alive with abundant plant life, bursting into bloom as if rejoicing. The wilderness, the burning sand, and the thirsty ground will gush forth pools, streams, and bubbling springs producing a lush oasis of plant life.

The LORD will surely comfort Zion...He will make her deserts like Eden, her waste lands like the garden of the LORD. (Isa. 51:3)

Four scripture references (Psalm 1:3, Jeremiah 17:7-8, Ezekiel 47:1-12, and Revelation 22:1-2) speak of a river of life flowing out of the throne room of God, beside which grow trees that never fail to bear their fruit and produce it in constant succession. Furthermore, the leaves of the trees possess medicinal and life-enhancing properties.

God is opening our eyes and our understanding in these pictures to a place of peace, rest, joy, companionship, shelter, provision, well-being, and meaningful and purposeful existence in his presence.

To him that overcomes, I will give the right to eat from the tree of life, which is in the paradise of God. (Rev. 2:7)

Even as God's Holy Spirit urged me, I invite you to be still and listen for his voice softly inviting you to inhale the fragrance of God's presence.

People who go through gardens without stopping to inhale the fragrance of flowers are getting their pleasure in one dimension only! The perfumes are like exquisite chords of music composed of many odor notes harmoniously blended.[8]

Miller

CHAPTER THREE

THE FRAGRANT TOUCH OF GOD'S SPIRIT

I.) The Anointing Oil

In the book of Exodus, we find a scriptural precedent using the language of fragrance. There, God ordained the scent of a holy anointing oil and a sacred incense to serve as significant reminders of his holiness and his presence, not only for his people but those in proximity to them.

In Chapter 30, God directs Moses to take the best spices, myrrh, fragrant cinnamon, fragrant cane (thought to be sweet calamus by plant scholars), cassia and olive oil and make a sacred anointing oil, blended together by an expert perfume-maker.

Table 1: Anointing Oil Ingredients

Spice	Botanical Name	Characteristics
Myrrh	*Commiphora myrrha*	A sticky gum-like resin collected in tears from small thorny trees native to Saudi Arabia; bitter tasting but when burned releases a powerfully fragrant odor; used in perfume making and as a salve that has healing properties; also used to anoint the dead; extremely costly[9]
Cinnamon	*Cinnamonum zeylanicum*	Strips of bark collected from 30' evergreen trees native to India; very fragrant; has antiseptic properties; used for flavoring; burned to release fragrance; very costly in biblical times[10]
Sweet Calamus	*Andropogon aromaticus*	An oil derived from cane crop grown in India; very sweet, spicy fragrance; highest quality oil comes from crushing part of the cane nearest to the ground[11]

Cassia	*Cinnamonum cassia*	Coarser and more pungent form of cinnamon; inferior to other cinnamon species; this spice often used to anoint the dead[12]
Olive Oil	*Olea europea*	Oil pressed from fruit of olive trees; most valuable agricultural crop grown in middle east; serves as a fixative, absorbs and enhances fragrance of other ingredients[13]

Essentially, scripture tells us this oil was a perfume. The word perfume comes from the Latin *per fumin*, "by means of smoke."[14] A glance at Table 1 reveals some measure of the fragrant and medicinal character of these spices. Most flowers, herbs, spices, and scented woods produce essential oils or resins that are the source of their fragrance.

During my days working in the nursery and greenhouse business, I often encouraged customers to touch the scented foliage of plants like garden mints and other herbs. The transfer of scent to their fingers by means of the plant's essential oil never failed to bring a look of amazement to their faces. The maximum release of many plants' fragrances is achieved by crushing or grinding the seeds, flowers, or foliage; by hanging them to dry; or by heating or cooking the various plant parts in question. Frequently, this process enhances the strength and character of the fragrance. Typically, although not always, the process of collecting and/or distilling the fragrance is very costly. For instance, attar of roses, a very concentrated fragrance made from the petals of damask rose varieties and cultivars, requires four thousand pounds of rose petals to make one pound of attar, which in turn sells for about $10,000! [15]

Several other characteristics of fragrance should be noted. It usually requires only small amounts of a fragrance to perfume a large area. The fragrance of one intensely perfumed rose can fill a whole room. Science has demonstrated that the fragrance of many plants serves to attract those that feed off of, or pollinate, them. Experts tell us that fragrance is the strongest sense tied to memory. Has a particular scent ever stirred a long forgotten memory of a person or place? Fragrance is also strongly tied to appetite. A stuffy nose or a cold often diminishes our enjoyment of a favorite food or meal. Lastly, fragrance provides many clues as to what is going on around us. Even if we cannot articulate why, we know when something smells clean or dirty, if something dead or decomposing is nearby, and in many cases, if something is poisonous. Not only is fragrance a functional warning system, it also serves as an adornment in the form of perfumes, colognes, and a whole host of other grooming products.

The anointing oil ordained by God to be used in tabernacle worship had several uses according to the Exodus 30 account.

> *Use it to anoint the tent of meeting, the ark of the testimony, the table and all its utensils, the lamp stand and all its accessories, the incense altar, the altar for burnt offerings and all its utensils, and the basin with its stand. (Ex. 30:26-28)*

Then two verses later,

> *Anoint Aaron and his sons . . . (v. 30)*

This anointing consecrated the tabernacle items, making them holy and acceptable to God. Not only were they to be subsequently associated with God's sacred nature, they were also to be set aside for a divine purpose, tabernacle worship. God's favor and presence rested upon everything that was anointed and also upon all who came into contact with these sacred objects. The same anointing on Aaron and his sons served to make them fit for godly service, thus enabling them as priests to carry out the atoning work that brought them into the holy presence of God without suffering the consequences of sinful man's exposure to the glory of God's righteousness. Its purpose and symbolism was so strong that the making and use of this fragrant oil came with a vehement warning from God:

> *This is to be my sacred anointing oil for generations to come. Do not pour it on men's bodies and do not make any oil with the same formula. It is sacred, and you are to consider it sacred. Whoever makes perfume like it and whoever puts it on anyone other than a priest must be cut off from his people.*
> *(Ex. 30:31-33)*

The practice of anointing with oil is found throughout the Bible. Although only olive oil was used, the spiritual significance remained constant all the way through New Testament times as a symbol of consecration, dedication, sanctification, and preparation. The depth and breadth of its significance can be seen throughout scripture.

Psalm 133 tells us,

> How good and pleasant it is when brothers live together in unity! It is like precious oil poured on the head, running down the beard, running down on Aaron's beard, down upon the collar of his robes. (v. 1, 2)

The anointing in this verse is a unifying outpouring. Much like the function of oil in the sacred anointing oil, it serves as a fixative, binding together, absorbing, and enhancing the properties of all the other additives in the blend.

Joined together, the body of Christ has a more extensive reach, a more powerful impact, and the capability to perfume or infiltrate a much larger space. Each vessel in the body helps to hold up and support the others. When one member is wounded or hurting, all join in sharing and diffusing the pain, thereby carrying some of the burden. Not only is the body more fitly joined together, but the anointing touch of God also draws us into a place of unity with the LORD our Healer to relieve our pain. With the LORD our Provider to meet our needs, the LORD our Comforter draws us into the shade of his presence, and the LORD our Peace stills our souls in the midst of the wind and the waves. The list of ways in which he has revealed himself goes on and on.

Scripture tells us that the oil of anointing, in the form of correction and redirection, restores us to right standing with God.

> Let a righteous man strike me—it is a kindness; let him rebuke me—it is oil on my head. My head will not refuse it. (Ps. 141:5)

Correction is never easy to receive, but when the righteousness of God is the standard by which our life is established, we must conform, for our own good, to the principles and precepts found in his word. Somewhere the psalmist discovered this spiritual principle. Before redirection becomes necessary, it is extremely beneficial to seek out correction, instruction, teaching, and direction. Get direction! Get teaching! Don't blindly set out on a course that will necessitate a painful change of direction. When we hunger and thirst after God's righteousness, life-directing words spoken in loving wisdom are indeed a blessing. They can lead us into a place of right standing that conforms to God's word and character.

Isaiah 61:3 tells us that God's anointing touch with the oil of gladness replaces the drab garments of sorrow and mourning. Imagine the effects of God's oil of gladness removing wrinkles of worry and creases of overwhelming concern. It changes our countenance and renews our youth! Its fragrance is a refreshing reminder of his favor.

In James, we are told to have the spiritual leaders of the church

> pray over [the sick] and anoint him with oil in the name of the Lord. And the prayer offered in faith will make the sick person well; the Lord will raise him up. If he has sinned, he will be forgiven. (Jam. 5:14, 15)

The oil here, as in the Old Testament, is symbolic of God's healing touch. It is also a reminder of the outflow of his Holy Spirit in the priesthood of believers. The gentle touch of the oil upon the sick is symbolic of the healing power of the Balm of Gilead poured out and into bodies and souls that are being tossed by the winds and waves of despair. If we read further, James writes,

> Pray for each other so that you may be healed. The prayer of the righteous man is powerful and effective. (Jam. 5:16)

In January of 2004 I lay in the intensive care unit of Emory University Hospital in Atlanta, Georgia. After several urgent phone calls made by my wife and family, news had spread through our various church families that my life was in grave condition. As I lay in a coma, four friends from my church drove the hundred miles late one night to Atlanta in rainy, wintry weather. A Christian nurse, despite the ICU rules concerning visiting hours, allowed them into my room where they anointed me with oil, lay hands on me, and prayed while many other faith-filled friends prayed in their homes. A few short hours after these prayers of faith were lifted up to heaven, I awoke. Despite three days of unconsciousness and the potential for multiple organ shutdown, I awoke in good enough medical condition to receive a liver transplant just two days later.

Lastly, the anointing oil is a type, a symbolic picture of the ministry of Jesus Christ as applied to those who reach out to him. In a later chapter, I will elaborate on how that ministry flows out of us, the body of Christ. All the previous types of anointing find their fulfillment in the ministry of Jesus Christ. The Hebrew word for Messiah, *mashiyach*, literally means "anointed" and comes from the word *mashach*, which means "to rub with oil."[16] Our word Christ comes from the Greek *Christos*, again literally meaning "anointed," and is derived from the word *chrio* meaning "to smear or rub with oil."[17]

This is my very basic understanding of how the ministry of Jesus Christ, the Anointed One, has touched my life and continues to impact the lives of countless others who turn to him. Jesus Christ, my Lord and Savior, came from glory to share in the weaknesses and temptations of my humanity, to live a perfect and spotless life in my stead, and to preach the good news that through him I might be set free from the bondage of sin. He came that I might be renewed and transformed into a new creation by his righteousness and that I might be restored to fellowship with God the Father, resting in his presence with all the benefits that are mine through him. Jesus Christ has placed his seal of ownership, his Holy Spirit, upon me in the form of a deposit, guaranteeing what is to come. Scorning the shame of the cross, he accepted what the Father laid before him even before they together created man. After laying down his life for me upon the cross and paying the penalty for my sin with his blood, he took it back up again, conquering sin and death. After rising from the dead, he ascended, returning once again to the glory that was his. There he now lives seated at the right hand of God the Father, interceding for me and presenting my needs and requests as my advocate. As I write, he is preparing a heavenly dwelling for me where I will live and reign with him for all of eternity!

The final test of Jesus' acceptance of the mission the Father set before him, and the powerful anointing that would enable him to ultimately fulfill that purpose, took place in the garden of Gethsemane. The garden was an olive grove, and its Greek name literally means "oil press."[18]

As his earthly ministry inched closer and closer to its final conclusion, the great weight of the infirmities, the sorrows, and the iniquities of all mankind grew heavier and heavier, pressing mercilessly upon his whole being. The overflowing abundance of the anointing upon him agonizingly spilled out upon the ground in great drops as he surrendered his will to his Father's plan. Thus, the prophecy of Isaiah was fulfilled,

> *Surely he took up our infirmities, and carried our sorrows. He was pierced for our transgressions, he was crushed for our iniquities; the punishment that brought us peace was upon him, and by his wounds we are healed...the LORD has laid on him the iniquity of us all. (Isa. 53:4-6)*

Clearly, the outpouring of God's anointing in our lives is incredibly significant, because it is part of the sanctifying work of God. Part of our growth as children of God involves dealing with the trials and tests that come our way. God knows we will run into terrible storms. Many of us, as I did, find ourselves in the midst of trying situations. We become exhausted, fearful, and frustrated physically, emotionally, and mentally by the storms we encounter. God knows that our natural response in the midst of uncertain circumstances is to fix our eyes upon the winds and waves of life or get so wrapped up in baling water that we forget that Jesus is in the boat! Like the disciples, we cry out,

> *Lord, save us! We are going to drown! (Matt. 8:25)*

Through his anointing, God means for us to be aware of his presence and his power to quiet the demons of fear, to calm our flesh, and to allow our spirits to focus on God, his word, and what is true even in the midst of the storm! The promise concerning the anointing that God placed on David's life in Psalm 89 is our heritage as well.

I have found David my servant; with my sacred oil I have anointed him. My hand will sustain him, surely my arm will strengthen him. No enemy will subject him to tribute; no wicked man will oppress him. I will crush his foes before him and strike down his adversaries. My faithful love will be with him and through my name his strength will be exalted...I will maintain my love to him forever, and my covenant with him will never fail. (v. 20-28)

Not only do we have a direct promise from God's Word, we can observe unique associations from the world of science that link fragrance to God's anointing touch. Just as fragrance is released by crushing or hanging plants to dry, so the power of God's anointing is released in us through the willing sacrifice of Jesus' death as he hung upon the cross. His fragrant touch consecrates us and prepares us for Godly service just as the sacred anointing oil consecrated the priests and the articles used in tabernacle worship. Just as the process of extracting fragrance is costly, so the fragrance of his touch increases our understanding of the extravagant cost of Jesus' sacrifice.

As fragrant flowers attract the birds and insects that feed from them, so his fragrant touch draws us into his presence to be nourished by his Word. Because fragrance is strongly tied to memory, his fragrant touch reminds us of past experiences of his deliverance, his saving grace, and his mercy. In the same way fragrance provides many clues as to what's going on around us, the power of his fragrant touch alerts us to the strategies and schemes aimed against us. In the same manner that oil acts as a fixative, blending and enhancing the properties of individual scents, so God's anointing oil joins us together with other believers in a spirit of unity. And lastly, just as leaves, resins, seeds and roots of many plants have medicinal benefits, so the gentle touch of God's anointing brings healing salve to the wounds of sin and rebellion in our lives.

II.) The Sacred Incense

Table 2: Sacred Incense Ingredients

Spice	Botanical Name	Characteristics
Myrrh	*Commiphora myrrha*	A sticky gum-like resin collected in tears from small, thorny trees native to Saudi Arabia; bitter tasting but when burned releases a powerfully fragrant odor; used in perfume making and as a salve that has healing properties; also used to anoint the dead; extremely costly[19]
Cassia	*Cinnamonum cassia*	Coarser and more pungent form of cinnamon; inferior to other cinnamon species; this spice often used to anoint the dead[20]
Spikenard	*Nardostachys grandiflora*	Gum-like resin collected from roots; imported from India; costliest perfume of N.T. times; used by Mary to anoint Jesus[21]

Saffron	*Crocus sativa*	Spice used for flavor, color, and fragrance; takes 4300 flowers to make one ounce of saffron, very costly[22]
Costus	*Saussurea lappa*	Perennial native to northern India; oil derived from the root; used as a an ingredient in perfumes[23]
Cinnamon	*Cinnamonum zeylanicum*	Strips of bark collected from evergreen trees native to India; very fragrant; has antiseptic properties, also used for flavoring; burned to release fragrance; very costly in biblical times[24]
Aromatic Bark	unspecified	unknown
Stacte	*Styrax officianalis*	Highly perfumed, resinous gum derived from the citrusy flowers of a stiff branched shrub native to the Middle East; used as a perfume[25]
Onchya	*Strombus lentiginous*	A fragrant essence thought to be derived from the inner lining of shells of various kinds of

		mollusks[26]
Galbanum	*Ferula galbaniflua*	Resinous, oily gum derived from the stems of a perennial native to Syria and Iraq; has a naturally bitter smell; used as a fixative because of its oily nature[27]
Frankincense	*Boswellia carterii, papyrifera, & thurifera*	Native to East Africa and Saudi Arabia; collected from large growing trees in resinous drops or pearls; purest form begins black but when burned turns to a milky white; long, slow burning; very sweet lingering fragrance; used for embalming ceremonies and fumigation[28]

In Exodus 30:34-38 God gave Moses the formula for a sacred incense to be used in tabernacle worship. The list includes an unspecified group of fragrant spices, followed by four specifically named spices, including stacte, onycha, galbanum, and pure frankincense. The Hebrew Talmud gives more details about the first group. It lists seven additional ingredients that were to be part of this blend: myrrh, cassia, spikenard, saffron, costus, cinnamon, and aromatic bark.[29]

From Table 2, it is easy to see that these spices were particularly chosen for their fragrant properties. Not only were they chosen for their fragrance, several of these spices were quite precious in value, including myrrh and cinnamon. Spikenard was considered the costliest perfume of New Testament times. The word from which galbanum is derived means "choicest part."[30]

The price of saffron remains very expensive to this day. Frankincense was so exclusive that it was associated with royalty, hence the present made to Jesus at his birth by the Magi.

This list of ingredients was to be blended together and ground into a fine powder by an expert perfumer. As with the anointing oil, they were not to make any of this incense for themselves. It was to be considered holy to the LORD and was only to be used by burning it as an act of worship inside the tent of meeting on a golden altar. As the fragrance was released, it was to be a reminder, a sign to Israel that the presence of God would indeed meet with them inside this tabernacle of worship.

Aaron burned the incense every morning and every evening. Perhaps most importantly, once a year on the Day of Atonement, Aaron was to burn the incense as part of God's covenant with Israel to forgive the sins of the people. He would first bathe and dress in priestly garments and offer the blood sacrifice. He would burn incense once again on the altar, directly in front of the curtain that hid the Ark of the Covenant. Then he would crawl around the side of the curtain, face and eyes looking down, and hang the fire pan that held the burning incense on one of the staves on either side of the Ark. As the fire pan swung, hanging from the staff, the burning incense would begin to create a cloud that filled the room where the Ark was located. Then Aaron would retreat the same way he entered, get the blood of the sacrifice, reenter and sprinkle the blood on the Ark for the atonement of his and his family's sins and for the sins of the people of Israel. The cloud had to cover the presence of God as it filled the room containing the Ark, or Aaron's life was in jeopardy. The cloud billowing above the tent of meeting served as a visual sign to all of Israel that God's presence was amongst them.

Perhaps an even more powerful reminder was found in the lingering fragrance of the burning incense. Not only did its perfume serve as a means of refreshing their awareness of him and their memory of his faithfulness, it covered the stench of the blood poured out on their behalf. Much later, as the temple was permanently established in Jerusalem, thousands of sacrifices were offered on the Day of Atonement. It is said that the fragrance was so powerful that it not only covered the smell of the sacrifices, but also perfumed the residents of Jericho forty miles away!

Just as the people of Israel were made aware of the presence of God with the lighting of the incense, so there are several additional things we should consider regarding the power of the fragrance of God's presence in our lives. One, it alerts us to his character. As God begins to reveal himself, we will feel the need to wash our hands, cleanse our hearts, and take off any filthy rags of unrighteousness that we may be clinging to. We innately know that God cannot and will not allow sin in his presence. As we confess our sins, God will begin to dress us in the garments of salvation. They are the white linen robes of the righteousness of Jesus Christ, having the same splendor as a bride and groom.

This principle explains why God wanted me to immerse myself in the fragrance of his presence. The voices of fear and the lies and schemes that the enemy was so desperately trying to bind me with could not follow me into God's presence. Not only is his presence a refuge, his character is also *a shelter from the storm and a shade from the heat. For the breath of the ruthless is like a storm driving against a wall and like the heat of the desert. You silence the uproar of foreigners; as heat is reduced by the shadow of a cloud, so the song of the ruthless is stilled.* (Isa. 25:4, 5)

The fragrance of God's presence, secondly, reminds us of how precious his presence is. Think of just one of the spices mentioned previously, spikenard, the costliest per-fume of New Testament times. Mary, the sister of Lazarus, according to the Gospel of John, Chapter 12, brought a jar containing about a pint of pure spikenard, broke it, poured it out on Jesus' feet, and then wiped his feet with her hair. Not only did the fragrance fill the whole house, this outpouring served as a preparatory anointing for his death and burial, an anointing that lingered with him all the way through his arrest, crucifixion and burial. The extravagance of this offering caused an uproar among those who held in higher esteem the monetary value of what was poured out (a year's wages) than the precious presence of their savior, Jesus Christ.

Consider the gold, frankincense, and myrrh presented to Jesus by the Magi near the time of his birth. These were gifts fit for a king. Consider some of the other costly and highly fragrant spices in the sacred incense, saffron and cinnamon. Consider the price paid by Jesus as he carried the sins of the whole world upon his shoulders. Consider the horrifying wounds and the brutal blows he endured, by which we are healed, and the unbearable weight of sorrows that was laid upon him. How precious the fragrance of God's presence is!

Thirdly, the power of the fragrance of God's presence can also be found by examining the Hebrew word for myrrh, *natah*. Literally, it means "drop" or "tear."[31] Over the years, its translation has been modified to mean "ooze" or "speak or act by inspiration."[32] When we enter the fragrance of God's presence, we begin to ooze with inspiration from God's Spirit. Our lives are perfumed by, permeated with, and infused by the fragrance of the Holy Spirit.

God's love, his joy, and his peace are produced as fruit in our lives. Instead of biting criticism, our mouths are filled with gentle, kind words. Instead of harsh and mean-spirited actions, our lives demonstrate God's gentle and kind-hearted spirit. We become aware of, and are filled with, God's goodness. We see the big picture well enough to enable us to continually wait, standing patiently and serving God and his purposes. We revere his faithfulness by becoming faithful ourselves. We are humbled by the example of Jesus who, in nature God, humbled himself, left glory, and came to us as a child. He worked and labored as a carpenter, refused a home, and laid down his life that he might become a faithful high priest and an ever-present advocate for those who would believe. Lastly, we are inspired to moderate our behavior to conform, to become like him. Jesus prayed in John's gospel, Chapter 17, for all those that ever came to believe, that they might become one with one another and also one with him and the Father. We become one with God by becoming like him.

Fourthly, we are placed under his roar, under his strong right arm, and under his protective care.

> The LORD will roar from Zion, and thunder from Jerusalem, the earth and the sky will tremble. But the LORD will be a refuge for his people, a stronghold for the people of Israel. (Joel 3:16)

The LORD will march out like a mighty man, like a warrior he will stir up his zeal; with a shout he will raise the battle cry and will triumph over his enemies. (Isa. 42:13)

The roar of his word goes out like a covering. His promises hover over us, shielding us from the enemy. Faithful and true, his shout protects us like a divinely crafted suit of armor.

An image from the book of Isaiah, Chapter 53, verse 10 paints a powerful picture of God's authority as a means of protection. Before all the nations of the earth, before even creation itself, God pushes up the sleeve of his garment and bares his right arm. In so doing, his strength, his power, and his great salvation are revealed to all. When we enter his presence, the enemies that would destroy us, that would steal from us, and that would strike us must flee. They cannot face the presence of an Almighty God who can overcome all by the truth of his Word and by the strength of his right arm.

Who can keep perfume on his hand from making itself known?
Proverbs 27:16 (Complete Jewish Bible)

Scent is what gives a rose its soul.[33]

John Fisher

Chapter Four

Adorned in the Fragrance of the Knowledge of Christ

What is the effect in our lives of the perfume of God's presence? Psalm 45, written by the sons of Korah, paints the first half of an intimate look at the answer to this question.

> Therefore God, your God has set you above your companions by anointing you with the oil of joy. All your robes are fragrant with myrrh and aloes and cassia. (v. 7)

Although this psalm was originally composed to honor David, King of Israel, the author of Hebrews informs us when he quotes these same verses in Hebrews, Chapter 1, that they also speak of Jesus Christ. The fragrance that the people of Israel clearly associated with God the Father in tabernacle worship accompanied Jesus as he descended from Glory to earth. This fragrant adornment caused him to stand out by revealing the presence of God to those whom he came in contact with.

The Apostle Paul furthers our understanding of the effect of this perfume in our lives, in 2 Corinthians, Chapter 2,

> *Thanks be to God (who always causes us to triumph) who always leads us in triumphal procession in Christ and through us spreads everywhere the fragrance of the knowledge of him. For we are to God the aroma of Christ among those who are being saved and those who are perishing. To the one we are the smell of death, to the other the fragrance of life. (v. 14-16)*

The perfume that adorned Jesus Christ as he ministered here on earth infuses us when we become born again children of God. This saving knowledge of Jesus becomes a fragrant adornment that causes us to become a message of life to some and an unnerving message of conviction to others. In a larger sense, it reveals the presence, the character, and the truth of God to those around us.

Just like the delightful fragrance of flowers prompts us to redirect our steps and to pursue the source of what caught our attention, so the fragrant presence of God draws those who are in need, who are perishing, towards the kingdom. It also fills those who are already saved with the desire to continually reenter the fullness of God's presence.

What is revealed in us when we are adorned with fragrance of the knowledge of Christ? What should we smell like? Is there any practical daily-living evidence of that fragrance? Jesus' life and ministry reveals a number of obvious concrete characteristics of the anointing that perfumes the lives of believers.

PURPOSE:

Purpose is one of the most distinctive elements revealed in Jesus' ministry. Scripture reveals that Jesus accepted the purpose God laid out for him before he and the Father created man. During the first five days of creation, God simply said, "Let there be," and by his word it came into being. Let there be light, let there be sky, let the waters be divided from the dry land and let there be seed-bearing plants and fruit-bearing trees. Let there be stars and heavenly bodies in the sky, let there be fish to fill the sea, let there be birds to fly in the air and let there be every kind of living creature upon the earth."

But on the sixth day, the account changes. As God the Father anticipates creating man, he includes his son, Jesus, in the decision. Implied in this statement, "Let us . . . shall we . . . are you ready to . . . make man in our image," is the tacit agreement that Jesus would follow through with everything that the creation of man entailed. "Are you ready to embrace your purpose? Are you ready to leave Glory?" In following through with the creation of man, Jesus gives his assent, "Yes, let us make man!"

In two Gospel of John accounts, Jesus clarifies any questions regarding his purpose here on earth. In Chapter 4, Verse 34 we read these words: "My food is to do the will of Him who sent me and to finish His work." Later in chapter 5, verse 17, Jesus says, "My Father is always at work to this very day, and I too am working." His sole purpose was to complete everything the Father had set before him to accomplish. Even though the "cup" he was asked to drink was horrifyingly bitter, and even though an unbearable burden of anguish was pressing down upon him, Jesus lifted up his voice for all of creation to hear and cried out, "Not my will, but yours be done." (Luke 22:44)

Jesus came with a purpose, he fulfilled that purpose, and even now he continues to serve an eternal purpose as our faithful High Priest. The embracing of that larger purpose brought a fragrant anointing to the life of Jesus Christ and should also adorn the lives of believers. God created us with a purpose in mind. We are part of a family that has a larger purpose.

During the worst of my illness, when life seemed to be slipping away and the daily challenge of facing the realities of terminal liver disease became more and more difficult, I admit that finding purpose in my life was extremely difficult. Even so, I discovered that I could hold on to one promise that a dearly trusted friend shared with me. He told me he didn't believe God was going to let me die at that time because he was sure God wasn't finished with me yet.

Although I struggled to find a purpose in what I was being asked to endure, I could hold on to this word of encouragement. It became for a brief season the only thing I could embrace and find meaning in. After I received the life-saving gift of a new liver, I was interviewed by our local newspaper, and I shared with the reporter all the things that my wife and family and I had held on to. Several days later, when the paper reported my story, it came out with this front-page headline, "God isn't finished with me yet!" That simple, uncomplicated statement became a beacon of light that reached into the community in a way that I would never have envisioned.

Patrick Morley, in his book *The Man in the Mirror* shares some keen insight into the significance of purpose in our lives.

> Purpose . . . answers life's larger questions . . . To be satisfying, our purpose needs to reflect our examination of life's larger meaning . . . Purpose is what God wants us to do long-term . . . [It acts] like a thread of continuity that is woven into the long-term view of our life. . . Purpose focuses and gives life direction . . . so that goals achieved do not become an unrelated string of hollow victories.[34]

The "why" of our existence is found in God's purpose for our lives. We have a mission, a destiny. The purpose that has significance and meaning, that will survive the test of time, is the one that is linked to God. Once we understand the direction He wants us to take, then we can begin to move in that direction.

If you have not discovered God's larger individual purpose for your life, you are not alone. Despite this, you can focus on the universal purposes laid out before all believers. Glorify God and enjoy Him—*purpose*. You are the salt of the earth—*purpose*. Serve the kingdom of God—*purpose*. As you go, make disciples—*purpose*. Walk in your gifts, take your one talent and turn it into two—*purpose*. Chances are, serving universal kingdom-wide purposes will lead you to a place and a time when you begin to walk in your individual purpose. The same fragrance that adorned Jesus Christ because of his faithfulness in serving the Father will perfume us when we dedicate ourselves to following God's purposeful leadership and direction in our lives.

FOCUS:

Another distinctive element revealed in Jesus' ministry can be seen in the way his eyes were constantly focused on his Father. This persistent attention accounts for why Jesus was able to pursue the purpose His Father had given him. As he gazed like the psalmist David upon the beauty of the LORD, his Father's countenance encouraged and sustained him when his flesh cried out with fatigue and hunger while walking in the path laid before him. Over and over, his eyes sought those of his Father like one utterly dependent. The smile of approval that he saw upon his Father's face revived him as it resounded over and over again with the words,

> *This is my beloved son, in whom I am well pleased.*
> *(Matt. 3:17)*

A scene from the movie *Jesus of Nazareth* vividly and powerfully portrays this dependence as Jesus is dying upon the cross. In the midst of unbearable physical suffering and with the weight of the whole world's sin pressing down upon him, Jesus' eyes turn to heaven, once again seeking the strength and sustenance found in his Father's countenance. For the first time, he finds his Father's eyes momentarily turned aside, and in great agony he cries out,

> *My God, my God, why have you deserted me? (Matt.*
> *27:46 CJB)*

Jesus set an example for us to follow if we desire to live in the fragrance of the knowledge of Christ. Jesus' eyes, as a result of being focused on the Father, were also focused on what is true. God's character and presence reveal what is true like a brilliant light. The enemy is knocked backwards and must retreat in the face of what is true. Lies and schemes viciously repeated over and over by the forces of darkness are silenced by the light of God's truth.

When the enemy's voice tried to convince me that God had abandoned me and was going to simply allow my health to waste away, God's spirit led me to this truth from his word.

In returning to me and resting in me, you shall be saved, in quietness and in trusting confidence shall be your strength. (Isa. 30:15)

I began to faithfully follow God's admonition. Hope began to replace my fears and uncertainties. Quiet confidence and trust in my God and in what is true led me into a place of peace and strength that enabled me to endure what was to come in just a few short months.

I was also challenged to focus on what I really believed and to remember what the word of God says he has done and will do for us. When we take up residence in the secret place of the Most High, are we truly protected from our enemies? Is his shelter a place of refuge and stability? Are we indeed covered by his precious promises? Is his truth our shield and protection as the psalmist writes in Psalm 91?

These principles and many others became my focus rather than the visible signs of the disease that was robbing me of life. I used this paraphrase of Psalm 30 to revive my soul, my spirit, and my flesh.

As I cry out to you and exalt you, silence the boasts and the taunts of my enemies, heal me, and restore my health. Lift me up from despair and sorrow; let your favor glow in my life. Despite the sadness of trials that visit and briefly plunge me into darkness, awaken me with the light of your glory. Remove the garments of suffering and let me be wrapped in celebration and in dancing. Let my well-being be a roar of thanksgiving filled praise and joyful shouting that can never be silenced![35]

Focus on what is true. Focus on what you believe, what you know to be true. Talk about what God has already done in your life and what his word says he will do for you. Then you will become a person that is not easily shaken. You will see beyond the now and your present circumstances. You will be held steady by God's truth rather than blown to and fro by the winds and waves of life. You will not be torn apart by delay. You will take joy in waiting upon your LORD because your expectation is in him. You will be renewed and strengthened, even in the midst of your weaknesses.

SERVANTHOOD:

A third principle is found in Jesus' assumption of the duties and the posture of a servant. He walked out of glory, voluntarily coming to earth and embracing the humble mantle of servanthood. A passage from the letter to the Philippian church explains how the example of Christ should be a model for all of us to emulate.

If you have any encouragement from being united with Christ, if any comfort from his love, if any fellowship with the Spirit, if any tenderness and compassion, then make my joy complete by being likeminded, having the same love, being one in spirit and purpose. Do nothing out of selfish ambition or vain conceit, but in humility consider others better than yourselves. Each of you should look not only to your own interest but also to the interest of others. Your attitude should be the same as that of Jesus Christ: who, being in very nature God did not consider equality with God something to be grasped, but made himself nothing, taking the very nature of a servant, being made in human likeness. And being found in appearance as a man, he humbled himself and became obedient to death, even death on a cross! (Phil. 2:1-8)

One of the keys to successful servanthood lies in making a choice, even as Jesus did, to look to the interests of others. The magnitude of this offering made on humanity's behalf was and continues to be staggering. His purpose required him to live as a man, to experience the temptations we are subject to and to face all the weaknesses common to mankind, without ever once succumbing to or giving in to temptation and without falling or giving in to weakness. Upon his success rested the salvation and redemption of all those who had and would yet believe. Even so, our service requires us to choose to humbly and willingly, without arguing or complaining, serve the purpose God lays before us.

The second key involves a decision based upon an understanding of the authority of Almighty God, the sovereign creator and ruler of all things. We must choose to honor that very same God with our service. Oswald Chambers writes,

There can be no moral virtue in obedience, unless a person recognizes the higher authority of the one giving orders.... A person is simply a slave for obeying, unless behind his obedience is the recognition of a Holy God.[36]

Successful servanthood understands and practices the honor principle. Not only should a servant look to the interest of others, but he or she should also value that service as a means of honoring God rather than as a choice rooted in ambition or any other form of prideful self-promotion.

The following illustration clarifies the significance of the honor principle. Suppose a stranger pulls a wadded up, filthy, torn piece of paper out of his pocket and offers it to the first person who will claim it. Those near enough to recognize various clues regarding its appearance would rush to claim that $100 bill. Despite some of the unpleasant characteristics of the soiled piece of paper, we recognize and appreciate its value enough to overlook any initial reluctance we might have to take hold of it.

Even as the Father exalted Jesus as a result of his determination to serve his Father's purpose, so God will cause the fragrance of the knowledge of Christ to perfume our lives when we choose to take hold of and honor him with service modeled after the example of Jesus.

ARRAYED IN THE GLORY OF RIGHTEOUSNESS:

The presence of God is an adornment that draws attention to and is a revelation of the reality of God and his character. The fourth principle found in Jesus' ministry is derived from God's character as it is revealed in Jesus' life, and in particular, the fragrant glory of his righteousness.

Many of us struggle with the concept that the righteousness of Jesus Christ is a mantle that God places upon those who are born again into the kingdom of God. When we look in the mirror, we don't see ourselves as God sees us. Rather, we see what we perceive to be true about ourselves. We have faults, we lose our temper, we procrastinate, we have lustful thoughts, and we open doors to things that we know we should avoid. We get caught up in our struggles instead of relying on and hoping in God.

Perhaps you grew up in an environment where you were constantly criticized and told that nothing you did would amount to anything. Regrettably, many bear the scars of emotional, sexual, and other forms of physical abuse. Perhaps no one ever spoke of their love for you or expressed their love to you in any real and meaningful way. Many struggle due to a wide array of totally absent forms of loving nurture and approval in their lives.

A long shadow hovers over the lives of people who have struggled in these areas, even if they have received biblical teaching regarding God's love. When they look in the mirror, that shadow often veils their understanding of the fragrance that Jesus' righteousness brings to their lives. Romans 3:10 tells us that, indeed, no one has lived up to the standard of God's righteousness. Isaiah 64:6 states that our own efforts—our own seeking after righteousness—only clothes us in the filthy rags of unrighteousness. But in Romans, Chapter 1, the Word of God says,

> *For man there is a righteousness now that is from God. (Rom. 1:13)*

Three times in the King James translation of scripture, the Word reads, "The just shall live by [his] faith." (Gal. 3:11, Rom.

1:17, Hab. 2:4 KJV) The NIV replaces the word "just" with the word "righteous." When German monk Martin Luther came upon this scriptural principle, his life was so heavily impacted that the Protestant Reformation was launched into being. While the church in Rome was selling indulgences (articles and certificates that guaranteed the buyer forgiveness for sins), Martin Luther's protest stubbornly proclaimed that righteousness could only be found in Christ.

> *When we are transformed, when our lives are made new by Jesus Christ, we become . . . endued with, viewed as in and examples of the righteousness of God—what we ought to be, approved and acceptable and in right relationship with Him, by his goodness. (2 Cor. 5:21, AMP)*

Although the righteousness of God is not birthed in us by our own works, we are admonished in Ephesians

> *to put on the new nature of true righteousness and holiness. (Eph. 4:24, AMP)*

In other words, now that you have been made right, begin to do what is right. God wants to array us in his righteousness so that we have something to give to others. We do that by immersing ourselves in Christ. We enter into the fragrant presence of God in any of a number of ways: through worship, through praise, through prayer, and through reading and meditating on God's Word. When we do, the aroma of his presence perfumes our lives. As we formerly read in Psalm 45, our garments (our lives) become fragrant with myrrh and aloes and cassia as our lives are anointed with the oil of joy.

David writes in Psalm 27,

> *One thing I require and that will I stubbornly insist upon, and that is that I might dwell in the house of the Lord forever.* (v. 4, AMP)

David longed for God's presence because he recognized that the desirability and the beauty of residing in and under the favor of God's countenance was an incredibly unspeakable place of blessing. He also yearned to embrace the character and the will of God while dwelling in such intimate proximity.

Paul reiterates in Philippians 2:12-15 that the fullness of God present in our lives creates in us a desire "to will and to do" his good purposes. When we not only desire his will, but actively begin seeking to fulfill his purposes, we put on true righteousness. When that righteousness from God takes hold of our lives, it begins to purify us and make us blameless children of God. It goes on to say in that passage that we subsequently begin to shine like stars in the universe, holding out and offering a hope to all we come into contact with.

2 Corinthians 4:6 tells us that the light of the knowledge of the glory of God shines out of darkness. Similarly, Psalm 37:6 asserts that the LORD will make our righteousness shine like the dawn, a time of day when darkness is replaced with light. These biblical principles remind us that the fragrance of the knowledge of Christ is being spread through us. Despite my faults, my inadequacies, and the sin that creeps into my life, in Christ I am a star that shines in the midst of darkness. Although darkness exists all around us, if just one glimmer of the light of the righteousness of God breaks through to someone we come into contact with who is lost or is struggling to survive in the midst of

violent and devastating circumstances, it is sufficient evidence of the sure existence of a Sovereign and Almighty God.

Those who are reflecting his glory according to 2 Corinthians 3:18, are continually being transformed into His likeness. In other words, people begin to become aware of and understand the presence and character of God and his glory. Those we work around, that we associate with, and that we live amongst become aware of God. We become a beacon, a light shining out of darkness, set upon a hill.

Like an exquisitely formed, freshly cut rose draws our attention, demanding that we pause to savor its perfume, so that fragrant, shining glimpse of the presence of God draws those who are hungry and thirsty, helpless, and needy towards the Savior. Unfortunately, it is also true that to those who are in rebellion, running from God and denying him, having embraced the "delights" of Satan's banqueting table, his presence has the odor of conviction, accountability, judgment and death.

Reconsider the picture you see when you view yourself. Is it possible for us to see ourselves in the light of his glorious righteousness, or does only God have the right to recognize in us this righteousness that comes from him? Scripture tells us that God expects others to see his righteousness in us. We may be only momentarily aware of the way his righteousness shines in us, but during times of trial, hardship, or great need; when our faith and trust prevail and we are moved to decision and action; that is when we put on his character and righteousness and reflect the glory of our Savior, our Healer, our Provider, our Comforter and the God who rules and reigns over all of creation.

THE MINISTRY OF RECONCILIATION:

The fifth area of Jesus' ministry that reflects how the fragrance of the knowledge of Christ is spread in our lives is the exhaustive ministry of reconciliation that Jesus embraced. From 2 Corinthians, Chapter 5 we read,

> *From now on we regard no one from a worldly point of view. Though we once regarded Christ in this way, we do so no longer. Therefore, if anyone is in Christ, he is a new creation. The old has gone. The new has come. All this is from God who reconciled us to Himself through Christ and gave us the ministry of reconciliation. That God was reconciling the world to Himself in Christ, not counting men's sins against them. And he has committed to us the message of reconciliation. We are therefore Christ's ambassadors as though God were making His appeal through us. (v. 16-20)*

Many church ministries focus principally on the verbal aspect of this ambassadorship. When I was in high school I was asked to participate in a community-wide effort to distribute a free book entitled *Right with God* to every home surrounding our church. Needless to say, I was apprehensive. As a young man I was unable, based on the overwhelming rejection my efforts met with, to appreciate the fact that many homes accepted this well-written message of the good news of Jesus Christ. Although many are gifted to approach people in this fashion and are very effective in presenting the gospel in this manner, I felt like a failure.

When we look at Jesus' ministry of reconciliation, we find that although there is an obvious and significant verbal component to it in the form of spirit-led preaching, teaching, and personal testimony, the ministry is not just verbal. It is also action based, need based, and servant based. It has purpose and a goal-oriented focus. This ministry's origins were defined by God long before Jesus set an example for us here on earth. In Psalm 146, the psalmist writes,

> *Blessed is he whose help is the God of Jacob, whose hope is in the LORD his God, maker of heaven and, the sea and everything in them. He upholds the cause of the oppressed and gives food to the hungry. The LORD sets prisoners free. The LORD gives sight to the blind. The LORD lifts up those who are bowed down. The LORD loves the righteous. The LORD watches over the stranger. The LORD sustains the fatherless and the widow, but he frustrates the ways of the wicked. (v. 5-9)*

In Psalm 146, we find that it is God's will for his people to stretch out their hands to the oppressed, to reach out to the fatherless, to touch the lives of those afflicted with sickness and sorrow. After seeing this ministry clearly defined, we find in Isaiah 61 that it has a personal component that Jesus echoed hundreds of years later and has subsequently been passed on to all believers.

> *The Spirit of the sovereign LORD is upon me because he has anointed me to preach good news to the poor. He has sent me to bind up the broken hearted, to proclaim freedom for the captives and release for the prisoners, sight for those held in darkness. To proclaim the year of the LORD's favor and the day of vengeance of our God. To comfort all who mourn, to provide for*

those who grieve in Zion. To bestow on them a crown of beauty instead of ashes, the oil of gladness instead of mourning, and a garment of praise instead of a spirit of despair. They will be called oaks of righteousness, a planting of the LORD for the display of His splendor. (v. 1-3)

Jesus fulfilled this ministry. The anointing of the LORD was upon him to touch people's lives not only with a message, but also a ministry of reconciliation. Ministry is another word for service. Service is another word for worship.

Scripture admonishes us to embrace this ministry, to accept this commissioning, to willingly consider it a privilege to serve as ambassadors "as though God were making his appeal through us." (2 Cor. 5:20) The Gospel of John reveals not only how Jesus embraced this ministry, but how we can follow Jesus' example, spreading the fragrance of God's appeal. In John, Chapter 3, Jesus revealed the great breadth and depth, as well as the divine power of God's love.

For God so loved the world that He gave His only begotten son, that whoever believes in Him should not perish but have everlasting life. (v. 16)

That message was spoken to someone who was experiencing a keenly felt spiritual emptiness in his life. This individual, whom we know as Nicodemus, surprisingly was a Pharisee, a spiritual leader, and someone who had been given the very best education in the written word of God. Pharisees were great students of the law and even wore sections of it strapped to their arms and foreheads. Despite this great "advantage," his spiritual hunger drove him to approach Jesus. Instead of a spiritual debate or an intellectual discussion, Jesus pours out a simple but incredibly powerful testimony that led Nicodemus to a new understanding of how God's love and his presence fills those empty places that we vainly try to fill with possessions, knowledge, relationships, power and influence. Nicodemus' struggle was put to rest in the fragrant aroma of the love of God.

Jesus reveals the approachableness of God in John, Chapter 4. He talked to a woman of Samaritan heritage, and he had answers. He had direction and he had good news for someone who hid her guilt and shame behind a conversation regarding the tedium of daily having to draw water. He revealed his insight into the problems of this woman's life by hinting that he had "living" water that would satisfy the real need she felt in her life. So thoroughly had Jesus laid his finger upon the problems in her that she left her water-pot by the well, ran back to town, and told everyone about the man she had met who knew everything about her.

When the disciples returned to the well after their search for food, they were shocked to see their master talking to a Samaritan, a people that the Jews despised, and a woman, which was culturally unacceptable in that day. Wisely, they kept their mouths shut. He refused their offer of food saying that he was "full." Sensing their confusion, he began his explanation of his actions that day by admonishing them to look around and notice the vast fields of human souls that were ready for harvest. He wanted them to see the great need that existed all around them. He wanted them to realize that anyone who is thirsty and needy, who is struggling with guilt and shame, may approach the one who has answers to all their needs. We can talk to God. We can ask God questions. We can, with the expectation of hope fulfilled, receive comfort and direction. Our God is a God who has good news.

In John, Chapter 5, Jesus revealed God's compassion for the infirmed and the sick. While visiting Jerusalem, Jesus came to the pool of Salome, a place where large crowds of people with all manner of sicknesses and physical handicaps gathered to enter the pool when an angel of the Lord stirred the waters. Among the crowd was a crippled man who had been ill for thirty-eight years. His handicap prevented him from entering the pool in time to receive healing. But Jesus was moved with compassion, knowing something of what was in this man's heart. He commanded him to roll up his mat, to stand and go home!

Numerous times Jesus reached out and touched the sick with, not only love, but with God's healing power. Scripture admonishes us to anoint with oil, to lay hands on the sick, and with the gift of faith pray in Jesus' name for their healing. When someone is ill and struggling with physically debilitating symptoms, they often wrestle with emotional and mental obstacles that can serve to hinder their ability to approach the throne room of grace. In these circumstances, God gifts others with generous portions of faith so that they can usher that individual past every hurdle into the healer's presence.

Consider the four who made their way to Atlanta to pray for me. Was it their faith, was it their fervent, effectual prayer, or was it their sacrifice? Without knowing a definitive answer to that question, we must simply follow the example of the healer, Jesus Christ. His compassion, his obedience to the purpose set before him, and his desire to rescue moved him to reach out and to act in the authority that was his. So God has given us this responsibility, has commissioned us to minister his loving grace to lay hands on the sick and to petition the courts of our God in the name of Jesus Christ to heal and to give people the strength to endure.

Jesus revealed, in the sixth chapter, God's desire to feed the hungry and to meet the needs of those willing to come into his presence. A multitude, hungry for more of Jesus' teaching and curious about reports that he was healing the sick, had followed Jesus and the disciples to a secluded hilltop. Already aware of their spiritual and physical hunger, Jesus asked the astonished disciples to bring him enough food to feed the enormous group of people gathered there on the hillside. After being given a paltry five loaves of bread and a couple of fish, Jesus commanded the hungry crowd to be seated, whereupon he blessed the food and began to break it up into pieces to be passed among the people. Imagine the reaction of both the disciples and the multitude as they realized that not one person went without among the thousands gathered that day!

Many people missed the point of the miracle they witnessed that day. Even today, despite the availability of study bibles, biblical preaching and Christian books authored by wonderful men and women of God, we miss the point. We think that "free bread" will solve our problems and satisfy our hunger. What God wanted those who were hungry to get that day wasn't just a meal. What he desires to honor is the hunger that leads us to follow him, to surround him, and to sit at his feet. Jesus promised that no one who comes to him would ever leave hungry. Yes, he will give us our daily bread, but there is a larger benefit to be found in his presence. Author John L. Mason puts it very well in his book *An Enemy Called Average* when he writes,

> It is more valuable to seek God's presence, than to seek his presents.[37]

Our emptiness will be satisfied and filled to overflowing in the bountiful fragrance of his presence. In the eighth chapter of John, Jesus revealed God's grace and mercy by ministering forgiveness and by setting people free from their bondage to chains of darkness. In this chapter, the religious leaders of the temple forcefully brought a woman caught red-handed in an adulterous relationship before him, thinking they would be able to trap Jesus into making some kind of error regarding the spiritual laws of Israel. As they questioned him, he paused and began writing in the dust around him. Perhaps his actions were in fulfillment of Jeremiah's prophecy,

> *Those that turn away from you [God] will be written in the dust. (Jer. 17:13)*

Their persistence finally brought a response from Jesus. "Okay, stone her, but let the person who is without sin throw first." Despite their anger, despite the fact that they were already armed with condemnation and judgment, God's mercy shamed them into silence; his blinding righteousness led them into the very trap they had set for Jesus.

Not only did this ministry of mercy and grace silence the enemy, it had a powerfully positive effect as well. As this woman laid wounded and guilt ridden at Jesus' feet, she heard her sentence of death overturned. The fear and guilt of sin were suddenly replaced by an overwhelming sense of mercy and pardon, of rescue and salvation! The strongholds that led her to debase herself sexually and cheapen her life, the chains of darkness that had held her in bondage, were suddenly broken by the brilliant light of God's forgiveness and mercy. From the confinement of sin, she was delivered into the glorious openness and spaciousness of freedom and forgiveness. For this woman, whom tradition says became a faithful follower of Jesus, the prophetic words of Isaiah 61, fulfilled in Jesus' exhaustive ministry of reconciliation, became true.

Early in my life, God convicted me of several strongholds, hidden closets that I had chosen not to deal with either out of guilt or shameful refusal to acknowledge that I had a problem. Each area of darkness had entered my life during a time of deliberate rebellion against not only my parents, but against all the spiritual hedges and boundaries that my parents and grandparents, the Holy Spirit, and God's word had established in my life. God led me to an altar where I confessed my sin to him and cried out in despair, feeling absolutely powerless to fight an addiction that I knew grieved my Savior. Unaware that that was exactly where Jesus wanted me, I was wondrously and gloriously delivered and set free in that very moment.

While God's deliverance from sin and bondage is sometimes instantaneous and eye opening, we still on occasion, either through our own stubbornness or by God's design, face battles and confrontations with the forces of darkness in our journey toward spiritual renewal. Just when our shield of faith has deflected the enemy's latest violent attack, another lie, another scheme, another attempt to trample our lives into the ground slaps us in the face. In these situations, God often has a future purpose. In order for us to accomplish that purpose in our lives, we must face trials and tests until we successfully defeat our enemy's schemes. God's design for some is transformation into battle hardened, combat savvy warriors that not only overcome personal strongholds, but also equip and assist others to destroy the strongholds they face.

Most of us have encountered a persistent weed in our gardens or landscape plantings. At some point in time, a weed seed is blown or washed into a space dedicated to desirable plants. As it grows, we often merely pull up the top of the plant or what can be seen. If any root is left in the ground, it will reappear at a later time growing more vigorously than it did previously. Pretty soon there is a thriving colony of that weed mocking all the attention you give to watering and fertilizing the special plants you actually purchased to plant in your garden.

In order for us to be thoroughly equipped to walk in the path God has prepared for us, every last bit of that weed, that stronghold, must be violently and completely uprooted. We must approach these battles with the faith and the knowledge that with God's help,

> *I can break through any barrier; with my God I can scale any wall. . . [It is] God who girds me with strength and makes the path safe before me. My feet you made swift as the deer's; you have made me stand firm on the heights. You have trained my hands for battle and my arms to bend the heavy bow. You gave me your saving shield; you upheld me, trained me with care. You gave me freedom for my steps; my feet have never slipped. I pursued and overtook my foes, never turning back till they were slain. I smote them so that they could not rise; they fell beneath my feet. You girded me with strength for battle; you made my enemies fall beneath me, you made my foes take flight; those who hated me I destroyed. They cried, but there was no one to save them; they cried to the LORD, but in vain. I crushed them fine as dust before the wind; trod them down like dirt in the streets. (Ps. 18: 29, 33-42, THE PSALMS A New Translation)*

We must understand that in the context of such victories, God has through trial and test equipped us with knowledge and wisdom for others going through similar circumstances. Because we have run the course, we should be moved with compassion even as Jesus was. And being so moved, we might pour ourselves into their lives to help birth spiritual victory in their journey. The fragrance of that outpouring is a powerful witness of God's grace and mercy in our lives. We as God's ambassadors must allow the glory of that fragrance to be spread through us as we go to all the world.

In John, Chapter 11, Jesus revealed his understanding and his compassion for those that are in the midst of seasons of suffering and sorrow, and he demonstrated his capacity to lift them out of the depths by anointing them with the oil of gladness.

Mary and Martha had sent news that their brother, Lazarus, was on his deathbed. Rather than dropping everything and rushing to heal Lazarus, Jesus briefly delayed his return to their home in Bethany. Even though the sisters had the faith to believe that Jesus would heal their brother, Jesus wanted to impart an infinitely greater, life-changing measure of gladness and joy that not even his intimate friends could conceive of.

To Mary and Martha's great sorrow, Lazarus died before Jesus arrived at their home. Upon his arrival he grieved and wept with all who had joined Mary and Martha in their grief, and then he followed them to the place of Lazarus' burial. With a loud voice he commanded Lazarus to come out of the tomb, and in an instant the cries of grief, the weeping and wailing that arise out of deep sorrow, were replaced with songs of praise, shouts of glory, and glad celebration.

In the present, we rarely understand God's timing nor can we grasp his infinite and sovereign purposes. Our God is not confined to the realm of the present. He can see and move from the present to the future and back. His purposes are so vast and so all encompassing that we must set aside our desire to fully comprehend, to know "why." With the gift of faith, we must simply embrace what is unseen and unknown. We will not be left to fend for ourselves.

The Apostle Paul reminds us in Romans, Chapter 8 that there is nothing that can separate from his love those who love God and have yielded to his call. No present danger, no future hardship, no financial crisis, no disease or sickness, no earthly or spiritual enemy, and no wound that might be inflicted has any power to drag us out from underneath the hovering protection of his love and his faithful promises.

Because of the LORD's great love we are not consumed, for his compassions never fail. They are new every morning; great is your faithfulness.... The LORD is good to those whose hope is in him, to the one who seeks him; it is good to wait quietly for the salvation of the LORD. (Lam. 3:22, 23, 25)

In John, Chapter 12, Jesus reveals God's desire to be a companion, not just a far away God. While still in Bethany celebrating with Mary, Martha and Lazarus, he visits the home of a man referred to as Simon the leper. Although the main story usually recognized in this chapter concerns Mary's extravagant outpouring of love, another lesson regarding fellowship stands out. God shows us through Jesus' example, his desire to walk with us as a companion in our daily living. He is found not just in our church or in religious activities but in all things. If we invite him into our daily life, he will walk with us as a friend, as a comforter, as a support, as one who gives of himself, and as one who guides us in our decision-making, keeping us on the right path.

The world we live in does not know God this way. Their knowledge is hindered by all the spiritual darkness that surrounds us. Even as Jesus set an example for us, it is incumbent on us to spread this fragrant awareness, this good news message, to those we come in contact with.

Jesus reveals in John, Chapter 13 God's willingness to sacrifice, to minister and to spend of himself and his resources in our lives. Jesus takes a towel and wraps it around his waist. He fills a bowl with water and in an incredible gesture of sacrifice begins washing the disciples' feet. The LORD of all the universe, the creator of all things, saw their concerns and needs. He not only saw them, he acted upon them. Verse 2 tells us he showed them the full extent of his love. He reached out, not mindful of his status or the glory that was his, and he spent himself on their behalf.

As the good shepherd, he ventures out to rescue us and to pour himself into our lives. He mends our broken hearts; he lifts up those who have collapsed in despair. He rushes out to defeat our enemies, baring his strong right arm on our behalf. He reaches into his vast storehouse and fills our life with eternal gifts, an inheritance that is beyond value. He draws us into his presence, into his peace, into salvation and reconciliation, and he shelters us from the stinking hot breath of the enemy. So we must follow Jesus' example and as servants of the kingdom of God, give of ourselves to build up and strengthen the body of believers.

In the fourteenth through sixteenth chapters of John's gospel, Jesus reveals himself as a counselor and an advocate. As he spoke of his impending death, the disciples became confused and fearful. Had the last three years been a waste? How could they continue on without Jesus?

Even as he instructed the disciples, we must lay worry and fear at the feet of the Father. The Counselor, the Holy Spirit whom the father has sent in Jesus' name, will bring us peace and will enlighten us and remind us of his words and his commands. We are not left to live alone. God's spirit will accompany us, leading us and directing us. Because Jesus has returned to the Father, we are commissioned to testify, to spread the fragrant knowledge of Jesus, the Christ. As our counselor, the Holy Spirit will give us a testimony that will bring glory to Jesus and to the kingdom of God by speaking of what he has heard and making it known to us.

Jesus himself has commissioned us, as recorded both in the gospels of Matthew and Mark, to make disciples as we go into the world. The Apostle Paul, under the inspiration of the Holy Spirit, has given this ministry a title as we read at the beginning of this chapter in 2 Corinthians 5: we are ambassadors, official representatives, spokespersons, people who do the King's business.

This ministry is uniquely carried out by believers in Christ. My experience is personal, much in the same way my relationship with God is personal. If we as believers try to be somebody else, such as a leader we admire or hold in esteem, we are probably doomed to frustration and ineffectiveness. God has placed us within a certain sphere of influence and has designed us to be effective within that sphere. Romans, Chapter 12 reminds us that we have different gifts according to the grace given us. This is by God's design. We can conclude that the significance of our personal ministry of reconciliation is tied to this unique gifting. Each of us is moved to function, to minister, and to serve according to who we are and how God created us and gifted us. God's fragrant touch uniquely perfumes our lives. Through this unique design, God has committed us as a body of believers to ministries of reconciliation that reach every sphere and touch every life.

> *We are therefore Christ's ambassadors, as though God were making his appeal [to be reconciled, to have our sinful record wiped clean] through us. (2 Cor. 5:20)*

*The way we are enticed into the garden
and encouraged to pursue its experience
to the end is like the plot of a novel.
It is the thread on which the
whole story unfolds.*[38]

Joe Eck

Scent is never still.[39]

Rosemary Verey

Chapter Five

Stepping Stones that Lead to the Fragrance of God's Presence

During the eleven or so years that I worked as a landscape designer, I developed an interest in creating pathways through some of the more detailed and extensive gardens I was designing. I came across a local provider of all sorts of natural stone who walked me through her family's stone yard. She related the various names, such as Tennessee Blue Orchard, and described the unique characteristics of the incredible variety of stones they offered for sale, some with fossil imprints, some with quartz crystals that reflected sunlight, and a multitude of colors and textures. I knew I had found a resource that would make my garden creations even more interesting and appealing to my customers.

The practice of using stepping-stones as a means of providing solid, dry footing for the garden visitor dates back to the fifteenth century. Flat stones were placed to practically match the human gait. When used most effectively, stepping-stones should create a path that provokes a thoughtful response to one's surroundings. The stones should also be of natural materials, thus blending in with and drawing attention to the garden rather than to the constructed path.

I identified four personal rules that served as inspiration in the creative design phase of the paths. First, the stepping stone path had to have a visual component to it that caught your attention. The stones I chose for creating a path always had variety in color and shape while still possessing a unified overall appearance. Secondly, the path had to allow you to get where you wanted to go without meandering aimlessly. It had to be functional. Thirdly, I always created the path to direct the visitor to specific points of interest along the way. The feature of interest might possibly be a bench or a shrub that had fragrant flowers or a stunning floral display. The path might also lead to plants whose foliage was maroon or variegated in color, possessed contrasting texture, or had unusual architectural value. Lastly, the stones must be spaced at comfortable intervals and be level and solidly placed so that they would be easy to walk on.

It is all well and good to be aware of a path and to have some understanding of its destination; it's another thing all together to actually step out and follow the path. Even after initially beginning our journey through the garden, we sometimes pause on a stepping-stone to reflect or to appreciate a view. But sooner or later we have to take another step, which brings us to the spiritual application of stepping-stones in our journey to and through the garden.

I would like to identify five stepping-stones that lead to the fragrance of God's presence, although I'm sure we could name many more. If you find yourself like I did, in need of peace, refuge, rest, healing, and encouragement, I urge you to take a first step toward God, inhale the fragrance of his presence, and gaze upon his beauty. Then take another step, another, and another until you find yourself hidden under the shadow of Almighty God.

THE POWER OF GOD'S WORD

The first stepping-stone leads us to the Word of God. There are things that are true about God's Word that I'm sure many of us don't fully understand. I recently heard a preacher say that, even if you don't totally comprehend everything you read in scripture, God's Word is so powerful that reading it hides enough of it in your heart and understanding to be an effective preventative against falling into sin and deception.

Despite this truth, God doesn't want that to be the extent of its effectiveness in our lives. We must begin thinking of the Word as a weapon—a weapon that we not only polish and sharpen, but actually strap onto our side. Having practiced with it and sharpened our skills with it, we become familiar with its advantages, its power, its length and breadth, and its ability to cut and penetrate to stop the advance of our enemies.

During World War II, the German military bragged of a super weapon in the planning and development stage that, when operational, would bring England to its knees in unconditional surrender and defeat. The very idea brought fear, doubt, and thoughts of capitulation, especially to the citizens of London.

Those that are born-again believers in Jesus Christ have such a weapon in the Word of God. No person, no philosophy, no spiritual force has ever defeated it! Yes, tragedies happen. Yes, bad things happen to good people. But if we fall into the trap of doubting the truth and reliability of God's Word, Satan has kept us from taking the first step into the garden and has turned us off the path toward God's fragrant presence.

We must pick up this weapon and start backing the enemy down with it. Yes, we can use it as a defensive weapon, but it would be a mistake according to God's Word to use it only that way. We have a whole arsenal of armor that is effective defensively, including truth, righteousness, the newness of salvation, peace, and faith.

Ephesians 6:18 admonishes us to think of God's Word as a sword. A sword is razor sharp. Because it has been polished and sharpened, it shines brilliantly. It is often made of superior metals that have been tempered or hardened to resist breaking. As such, its ability to maim and destroy instills fear in those against whom it is wielded.

Despite its obvious power and effectiveness, the forces of darkness are still frequently going to advance against us. Satan hates God, those who have been reconciled to God, and all that reflects the glory of God, and he desires to deceive, corrupt, and destroy everything that is of God. Furthermore, he desires to frustrate the plan of God to reestablish fellowship with mankind in the heavenly fragrance of paradise.

> *For our struggle is not against flesh and blood, but against the rulers, against the authorities, against the powers of this dark world and against the spiritual forces of evil in the heavenly realms. (Eph. 6:12)*

Scripture is filled with warnings regarding the strategies of the enemy and the power of the Word to defeat these strategies.

> *Put on the full armor of God [including the sword of the Spirit] so that you can take your stand against the devil's schemes. (Eph. 6:11)*

> *The weapons we fight with are not the weapons of the world. On the contrary, they have divine power to demolish strongholds. We demolish arguments and every pretension that sets itself up against the knowledge of God. (2 Cor. 10:4, 5)*

One of the best examples of the divine power of God's Word is found in Matthew, Chapter 4. Jesus had been sent out into the wilderness by the Holy Spirit to prepare him for his earthly ministry. Satan came against him using three strategies. Although this account does not tell us that Jesus was directed to fast for forty days, we may presume that by fasting he was following the will of his Father. Knowing that Jesus was hungry, Satan first tempted him to satisfy his physical need. Clearly he wanted Jesus to abandon his reliance on his Father's provision and surrender to his fleshly appetite. Jesus responded by quoting from Deuteronomy,

> *It is written: 'Man does not live on bread alone, but on every word that comes from the mouth of God.' (Deut. 8:3)*

Defeated by the Word, Satan tried again. In so many words, he arrogantly suggested, "How about jumping off this wall? The Word says God's angels will protect you and keep you from hurting yourself. If God's Word is really true, let's see something spectacular." Sowing doubt regarding the truth of God's Word had worked well for him in the first Eden. But again, Jesus answered with the Word,

> *It is also written: 'Do not put the Lord your God to the test.' (Deut. 6:16)*

Lastly, the tempter took Jesus to a very high mountain and offered Jesus lordship over all the kingdoms of the world and their earthly splendor if only Jesus would bow down and worship him. Once again and in triumphant fashion, Jesus withdrew the two-edged sword of God's Word and commanded Satan to back off and flee.

> *Away from me, Satan! For it is written: 'Worship the Lord your God, and serve him only.' (Deut. 6:13)*

Jesus used the Word of God to demolish every one of the enemy's arguments, lies and schemes. Do we have enough Word hidden our hearts to use the sword of the Spirit like this? Or is it dusty and unfamiliar? When the enemy says, "You're going to fail; you're going to drown in sorrow; I'm going to take your life; you are meaningless; you can do this and still be a Christian," are we familiar enough with the Word to say, "Not according to my God!"?

In addition to the divine power of the Word of God to demolish every pretension that sets itself up against the knowledge of God, scripture also says that God's Word will bring things into being that are not seen, and will firmly establish them upon an unshakeable foundation. I like how the authors of the King James Version of the Bible translate Psalm 138:2. It gives us a definitive picture of how God regards the words that proceed out of his mouth:

> *For thou hast magnified thy word above all thy name. (Ps. 138:2b, KJV)*

The prophet Isaiah records on several occasions what God himself has said about the words he has spoken:

> *What I have said, that will I bring about; what I have planned, that will I do. (Isa. 46:11b)*

> *So is my word that goes out from my mouth: it will not return to me empty, but will accomplish what I desire and achieve the purpose for which I sent it. (Isa. 55:11)*

One particular scripture ministered to me tremendously and gave me hope in the midst of my illness:

> *[Then] He sent forth his word and healed them. (Ps. 107:20)*

Psalm 119 is probably the greatest chapter in the Bible when it comes to highlighting the benefits and the good things that come about as a result of God's Word. In a very abbreviated paraphrase, Psalm 119 tells us that by reading and obeying God's Word, his promises, and his laws, by meditating on them and using them as a standard, we will be:

BLESSED,	MADE PURE,
HEDGED,	PROTECTED FROM SIN,
COUNSELED,	RENEWED,
STRENGTHENED,	DELIGHTED,
SET FREE,	GIVEN HOPE,
COMFORTED,	DIRECTED,
SUSTAINED,	ENLIGHTENED

& FILLED WITH PEACE.

Jesus, whom we also know as "The Word of God" according to the Apostle John, also speaks of the power of the word throughout the course of his earthly ministry:

> *Therefore everyone who hears these words of mine and puts them into practice is like a wise man who built his house on the rock. (Matt. 7:24)*

Although the rain of difficult circumstance beats down upon us; although strong winds of adversity strive to bend and uproot us; and even though waves of seeming impossibility break over our heads, if we allow our lives to be established on the unshakeable foundation of the Word of God, we will not fall! In fact, if we embrace and apply the Word to our lives, allowing it to take root, Jesus said that it will produce an abundant harvest of fruit, a yield many times greater than what was sown. (Matt. 13:23)

THE PRIVILEGE OF PRAYER

Many learned individuals have written exhaustive works on the subject of prayer. This brief section represents what I've learned from the influential leadership of great pastors, from personal study, and from my own experience.

I believe a huge majority of believers are burdened by numerous misconceptions about prayer that result from some decidedly intimidating and legalistic rules that have arisen in the body of Christ down through the centuries. How long do we have to pray? Do our eyes have to be closed? Does it count more if we are on our knees? Can more than one person pray at the same time? What should prayer sound like? Should it be eloquent? I remember a unique prayer moment from my youth when a ten or eleven year old boy, newly adopted by a family in our church, launched into a frankly hilarious prayer spoken in his best King James English petitioning the courts of heaven for a sufficient number of disposable diapers for his brand new baby sister. I imagine God the Father had a gleam in his eye when he proclaimed to the host of heaven, "Done!"

I think there is little value in trying to lay down concrete guidelines about how long and where and when to pray and whether our prayers should be out loud or silent. Guidelines of this sort tend to launch us into the "performance track" of spirituality. Furthermore, I believe that doing so hinders honest, heartfelt prayer. Some of the debate regarding these issues falls into the realm of "blockhead controversies," part of Satan's strategy to create stumbling blocks that keep us from even getting started.

Many of the answers to the questions above should arise out of personal conviction and the nature of the prayer being lifted up to God. Prayers for deliverance might be accompanied by weeping and intense petitioning. Prayers of thanksgiving might be accompanied by shouts of rejoicing and dancing before the LORD. Prayers of dedication or consecration might be accompanied by solemn formality and predetermined liturgical responses. Prayers of repentance might be poured out while kneeling or prostrating oneself before God. Follow God's Spirit and allow prayer to arise out of the cry of your heart.

The Word does indeed give us some general guidelines about when to pray—anytime or all the time! David cried out "morning by morning" (Ps. 5:3), and at "evening, morning and noon (all day)." (Ps. 55:17) Paul writes in his final instructions to the Thessalonian church, "pray continually." (1 Thes. 5:17) In other words, there is no inappropriate time to pray. Call out to God when you need to unburden your heart or when you see a need.

If we look at the relationship God had with man in the Garden of Eden, we find the foundation of every kind of prayer. Mankind enjoyed the fellowship of communing with and being in the presence of God the Creator. With that in mind, we can define prayer as an earnest expression of love, appreciation, and need directed to God. It involves reflection, requests, declarations and petitions. While you and I no longer walk in God's presence like Adam and Eve did, prayer is the conduit, the built-in response of man to God that today establishes fellowship with our Creator.

As such, several significant principles about the nature of prayer are revealed. First, prayer acknowledges the reality of God. There is no reason to pray if there is no God. Secondly, prayer reveals that we have some basic awareness of who we are in relation to God. He is the Creator of all things and we are the created. Whether we fully comprehend it or not, the act of praying acknowledges God as the all-powerful, limitless provider and source of all that is good, and recognizes that his warehouse is inexhaustible and his reach is infinitely long.

Numerous prayers of different sorts are recorded in scripture, including prayers of intercession, prayers of consecration and dedication, corporate prayers of repentance, prayers for the sick, as well as examples of intimate communion with God. Rather than looking at them as concrete examples to follow (after all, they were offered in the context of very specific circumstances), they should be studied like blueprints for the principles illustrated in their content. Although the word "blueprint" might convey an association with exacting details, I would like to use the word more in the context of an artistic spirit-realm rendering of what God felt was necessary to be revealed regarding the subject of prayer.

The Psalms give us an outstanding blueprint to study. David often practiced a kind of prayer others have called the prayer of communion—prayer that leads us into a place of reflection and revelation. As he came before the LORD, he began by recognizing who God was and the ways in which he had shown himself faithful to his people and to himself. Then David would reflect upon his circumstances and his needs. His prayers always concluded with faith-filled declarations that proclaimed that God would rescue him and deliver him from the hand of his enemies regardless of his present situation and despite what appeared inevitable. One of my favorite psalms, Psalm 27, demonstrates this. David begins the psalm with a statement regarding the nature of God.

The LORD is my light and my salvation—whom shall I fear? The LORD is the stronghold of my life—of whom shall I be afraid? (v. 1)

Notice the salutation, the "Dear God" part of this blueprint. David addresses God by a name that was personally significant in his life and by a name that described and revealed God's character. It was a name that God recognized. David knew that this kind of acknowledgement of God was a key that unlocked doors and opened avenues of communication with God, thus allowing him to enter into God's presence.

The privilege of prayer allows you and me to address God by one of the many names that he has given us that reveal who he is. Do you know him as your Savior, as your Deliverer, as your Provider, as your Comforter, or as your Righteousness? Then call him by that name! Your recognition will reinforce your faith and trust in him and cause his glory to perfume your life.

Because of my experience with illness and the resulting despair, I often now recognize God as "The One Who Spared Me From Death," or as "The One Who Bared His Strong Right Arm on My Behalf." Although they may not be formal "names" of God given to us in scripture, they describe how I know him!

After this opening (in Psalm 27), David asks the LORD to be merciful, to teach, to lead and guide him even in the face of his enemy's evil desires and violent assaults. He pleads with God not to abandon him in this trying time. Despite feeling threatened by the false witnesses that had risen up against him, he concludes with this declaration concerning the character and nature of his God,

> *I am confident of this: I will see the goodness of the LORD in the land of the living. Wait for the LORD; be strong and take heart and wait for the LORD. (v. 13, 14)*

Jesus himself gave us a blueprint for prayer. Interestingly enough, I believe a significant principle is found in the fact that it is only four sentences long. Obviously, God does not have an egg timer on his children, recording whether they have met their prayer quota for the day. Nor does he have a checklist upon which he grades us for including every one of scripture's admonitions about prayer.

Jesus' blueprint, just like David's, begins with an acknowledgement of who God is, "Our Father in heaven." (Matt. 6:9) John 17 also records that when Jesus prayed in the upper room, he looked to heaven and cried out, "Father . . ." (John 17:1) Mark's gospel tells us that, just a few hours later while in the Garden of Gethsemane, Jesus cried out, "Abba, Father. . ." (Mark 14:36)

These prayer salutations are significant for several reasons to people seeking to commune with God. They remind us of our personal relationship with God and of our personal experience of him. The God of heaven and earth, who created all things and who is LORD of All, created us to walk with him and to enjoy the fragrance of his presence. Addressing him as Father reminds us of his unlimited ability to meet our three most critical needs, to love us, to shelter us, and to provide for our needs. Even though many of us have had loving fathers, our appreciation level raises higher when we consider that our God is an all-powerful, ever present, invincible, infinitely loving and eternal Father. Remembering that he dwells in heaven is also significant, because it acknowledges our earthly position in relation to his supreme authority and in relation to the exalted place of the Father's throne. Though Jesus was in very nature God (the image of the invisible God, as Paul identifies him in the letter to the Philippian church), his example of prayer testifies to us today that he humbled himself and made himself a servant to his Father's will as an example to us, God's people.

2 Chronicles 7:14-16 reveals that God will honor the person who prays humbly. Humility involves choosing to bow to God's will. It involves turning from choices and behaviors that lead to sin, and earnestly seeking God's face. He promises Solomon here that if his people would approach him in this fashion, he would focus his eye upon them, and his ears would be attentive to their prayers. When we bow our will before God, acknowledging our awareness that we were crafted from the soil of this earth, we open a door into his presence. Although in this passage God is specifically addressing prayers that would arise from the temple that Solomon had built as God's earthly dwelling place, he prophetically reveals an amazing promise to all those adopted into the household of God.

I have chosen and consecrated this temple so that my Name may be there forever. My eyes and my heart will always be there. (v. 16)

This humble approach enhances another significant principle of effective prayer. We must submit to the plans God has for us. Submission not only conveys a bending of our will, but also an embracing of God's will. Jesus' blueprint paints a vivid picture of this attitude:

Your will be done on earth as it is in heaven. (Matt. 6:10)

When God conveys his desires to the Host of Heaven, can you imagine any of the angels shuffling his feet and saying, "Do I have to?" The picture that comes to my mind is one of a million angelic faces looking at the Father with an eager expression of agreement, "Yes, Yes," and a barely contained attitude of anticipation, "Pick me, let it be me!" I don't see any of the angelic host lagging behind. Rather, I imagine the Host of Heaven stampeding to bring about God's purposes.

This attitude of submission does not mean that we are to surrender to our circumstances thinking that God has foreordained us to endure merciless trials, tribulations, poverty, and sickness. God's plan for us will involve some shaping and molding. We must go through an often-uncomfortable refining process that will turn us into a vessel useful for service in the Kingdom of God. By his power he will enable us to endure this process and emerge as shining examples of his glory. We must remember that although we are consumed by our physical and immediate needs, God sees a different picture. He sees how our present circumstances will have an eternal impact on the whole of the Kingdom of God.

Just as David did, Jesus tells us to bring our petitions and requests before the LORD.

Give us today our daily bread. (Matt. 5:11)

We are to do so with the understanding that God already knows our needs and that as the creator of all things he is also the generous provider of all things.

See how the lilies of the field grow. They do not labor or spin. Yet I tell you that not even Solomon in all his splendor was dressed like one of these. (Matt. 6:28, 29)

Even though God knows our needs, he clearly wants us to ask, thus acknowledging once again that it is through him that we are daily sustained. Our cries for rescue and healing, for provision and favor, are demonstrations of belief that only God can satisfy those needs and that our strength is not sufficient.

The Apostle John recorded further instructions given by Jesus about asking God for daily provision:

> *If you remain in me and my words remain in you, ask whatever you wish, and it will be given you. (John 15:7)*

Has God's Word been solidly planted in our hearts? Have we steadfastly remained established on the foundation that God has laid in our lives? Are the requests that proceed out of our mouths the fruit of spiritual deposits securely invested and cultivated in our lives, or are they the fruit of our own selfish desires? After all,

> *Out of the abundance of the heart the mouth speaks. (Matt. 12:34)*

This verse reminds us that God is listening to our requests with a sovereign understanding of our motives. With that in mind, we should once again be reminded that his presence is more valuable than are the presents we often desire.

Scripture has much more to say regarding presenting our requests before the LORD. Prayer that arises out of significant hunger and sincere need, and is poured out urgently and with deep fervor, touches God's heart.

> *The effectual, fervent prayer of a righteous man availeth much. (James 5:16 KJV)*

Effectual literally means "fully adequate."[40] Effectual prayers powerfully accomplish God's purposes. After enduring years of emotional pain as a result of being childless, Hannah poured her heart out to God in great anguish and grief. Not only did she petition God for a son, she declared that if her prayer were answered, her son would be dedicated to serving God for all the days of his life. The son that God blessed Hannah with was named Samuel, which means "heard of God." He played a significant historical and spiritual role in ushering in the kingdom of David, of whose lineage Jesus was born. Notice how powerfully God was moved by the fervent prayer of his servant Hannah.

God answered my own "earnest" prayer. My prayer was lifted up in real and literal desperation. I could feel death's mocking approach. I had allowed despair to weaken my confidence in God. God's powerful answer, "Be still and take a deep breath…" brought peace to my troubled mind and heart and silenced the enemy's mocking voice.

As we present our requests before the LORD, keep in mind that God sees us as his children. Jesus reminds us in Matthew 7 that even as we know how to give good gifts to our children, God, our spiritual Father, is infinitely more able to give even greater gifts to those who ask him! This promise is preceded by the command,

Ask and it will be given to you; seek and you will find; knock and the door will be opened to you. For everyone who asks receives; he who seeks finds; and to him who knocks, the door will be opened. (Matt. 7:7, 8)

How does a child ask? Over and over and over again until Daddy or Momma listen and grant his request. Jesus instructs us, "Keep on asking; keep on seeking; keep on knocking!"

A child also asks in another unique fashion. He or she asks without any doubt that his parents can provide. This is simple child-like faith. Children have no problem seeing things through faith. The writer of Hebrews tells us that faith is being sure of what we hope for and certain of what we can't see. He also reminds us that when we approach God with our desires, we must believe he exists and that he rewards those that earnestly seek him. (Heb. 11:1-6)

Like a child, we are to ask without doubting or wavering in our faith. Children don't do budget analyses; children don't ever believe their parents' resources are exhausted; and children don't look around and say, "Since Johnny didn't get a bike, I better not ask for one." They ask, believing in their parents' power to provide. Prayer is, as mentioned earlier in this discussion, an acknowledgement of God's power to reach into his storehouse and offer a blessing that is beyond our reach and our power to imagine. The Apostle Paul writes,

Your faith should not stand in the wisdom of men, but in the power of God. (1 Cor. 2:5 KJV)

After Jesus' direction to lift our needs up to God, he very briefly identifies another key that opens a door in our prayer of communion. He says,

Forgive us our debts as we forgive our debtors. (Matt. 6:12)

The reluctance to forgive or to forgo keeping track of the wrongdoing of others is perhaps one of the biggest hindrances in the prayer lives of believers. Christians who are clinging to the right to be offended by what others have done or what they believe has been done to them, to family, or to friends have taken a very dangerous spiritual stand. In one of the most important books authored in the recent past, *The Bait of Satan,* John Bevere writes, "Our response to an offense determines our future."[41]

Many of us desire forgiveness without accountability. Jesus illustrates the danger of this lack of accountability in one of his parables. In Matthew, Chapter 18, he relates the story of a man who, being found in great debt, is brought before his master to give an accounting. When it is discovered that he is unable to pay his debt, his master sentences him and his family to be sold into bondage. Upon hearing his sentence, he begs his master for forgiveness, which is granted, and the debt is erased. Shortly thereafter, the same man approaches someone who is in debt to him and, finding him unable to repay his debt, pronounces the very same sentence upon him for which he was forgiven a brief time before. When this is discovered, the first man is sentenced once again to a life of bondage. Jesus concludes the parable with these words,

> *This is how my heavenly Father will treat each of you unless you forgive your brother from your heart. (Matt. 18:35)*

In this light, John Bevere's words take on significant meaning. An ongoing, intimate relationship with God will be severely handicapped by refusal or reluctance to forgive. It is part of God's character to forgive. If we choose not to forgive, we are in effect judging God's character to be flawed. This raises a question of eternal importance regarding our future, "Who can you afford not to forgive?"

Jesus elaborates even further with these words,

> *With the measure you use, it will be measured to you. (Matt. 7:2b)*

This statement reveals something even bigger than the importance of forgiveness, leading us to the following principle. What we desire from God must be reinvested in the lives of others. If we desire compassion, we must be willing to be compassionate. If we desire peace, we must become peacemakers. If we desire provision, we must be willing to spend ourselves on behalf of others. If we desire God's healing touch, we must reach out to comfort others in the midst of their despair. I believe that God has not answered the prayers of many because they have not been faithful to follow this principle. Have we forgotten who we are? Have we forgotten how great our God is? Have we forgotten how great a debt we have been forgiven?

The last guideline that Jesus gives in his blueprint for prayer is to pray "in Jesus' name."

> *I will do whatever you ask in my name, so that the Son may bring glory to the Father. (John 14:13)*

> *My Father will give you whatever you ask in my name. Until now you have not asked for anything in my name. Ask and you will receive, and your joy will be complete. (John 16:24)*

As Jesus' earthly ministry was coming to a close, he added something to his blueprint. He reminded the disciples just before his ascension into heaven that all authority had been given to him by his Father. He had become the mediator, the intercessor, and advocate for all who would seek atonement from God. The writer of Hebrews tells us that after his ascension, he sat at the right hand of God the Father. (Heb. 1:3) There, he lives to intercede for those who come to God. Because he has shared in our humanity, he serves as a merciful, compassionate, and faithful High Priest on our behalf. He has seen our suffering not only with divine eyes, but also with the eyes of man.

> *Therefore, since we have a great high priest who has gone into heaven, Jesus the Son of God, let us hold firmly to the faith we profess. For we do not have a high priest who is unable to sympathize with our weaknesses, but we have one who has been tempted in every way, just as we are—yet was without sin. Let us then approach the throne of grace with confidence, so that we may receive mercy and find grace to help us in our time of need. (Heb. 4:14-16)*

Praying "in Jesus name" isn't just some trite ending tagged on to our prayers. It is an expression that brings us full circle to the place where we began, acknowledging God and who he is. Jesus, the Son of God, with whom we are heirs of a tremendous heritage, is now the means by which we confidently approach God's throne. As one who has seen and experienced our hunger and our passions, the temptations of self, the fatigue of daily living, and the rejection of enemies intent upon our destruction, we have a faithful, effectual (fully adequate) intermediary perfectly offering and wording our reflections, our requests, our declarations, and our petitions to our Father God, the Creator of All Things.

In summary, prayer is the act of putting declarations and reflections upon who God is into thought or words and humbly offering them anytime and at all times in reverence. In prayer, we convey our willingness to bow to God's will and to his way, even as we earnestly present our desires to him just like a son or daughter. We must acknowledge that whatever we petition the courts of heaven for must be reinvested into the lives of others. When we dedicate ourselves to communing with God in this fashion, our praying "in Jesus' name" becomes not as much an exercise in getting answers from God as it becomes finding that secret place of rest in the fragrant presence of God.

THE PLEASURE OF PRAISE

Let everything that has breath praise the LORD. (Ps. 150:6)

With these words, the psalmist concludes and sums up everything written in the whole of the Psalms. Not only is everything that has breath admonished to praise God, but God's Word goes on to say in Psalm 148 that all of creation is to praise his name. All his angels, the sun and moon, all the shining stars, the highest heavens, the great sea creatures, the ocean depths, lightning and hail, snow and clouds, stormy winds, mountains, hills and rocks, fruit trees, cedars and all the trees of the field, wild animals, flying birds, all nations, kings and princes and rulers, young men and young ladies, the elderly and children alike are to

praise the name of the LORD, for his name alone is exalted; his splendor is above the earth and the heavens. (Ps. 148:13)

What is praise? According to the great 19th century preacher and writer C. H. Spurgeon, praise is the "intelligent admiration of God, kindled into flame by gratitude and fanned by delight and joy."[42] Praise is an acknowledgement of who God is and what he has done. It is a compliment! Praise should be spoken publicly, individually and collectively. It should be declared openly, plainly, clearly, and joyously with boasts and shouts, with lifted hands, with song, with musical instruments, with hands clapping, with feet dancing, with rejoicing, in triumph and with adoration. Even the world gets this!

Acknowledging that there must be times of reverence and silence, I wonder if more church congregations ought not to resemble the crowds at football games rather than the stale, boring, unappreciative, and disrespectful places that some have become. After all, God's Word tells us that praise is the accepted norm of the entire universe! It is where God lives and is enthroned. Praise is heart worship.

In the Old Testament, there are seven Hebrew words that are translated into the English word "praise." In order to clarify what God's Word has to say about acceptable praise, I will briefly list and define them here.

Table 3: The Seven Hebrew Words for Praise[43]

Hebrew	Definition	Bible Ref.
Halal	to boast, to celebrate exuberantly, an explosion of enthusiasm connected to the overthrow of an enemy	Psalms 22:22, 63:5, 117:1

Yadah	to thank by public acknowledgement, to extend the hand	II Chron. 20:19-21, Psalms 9:1, 134:2
Barak	to bless, to bow, to kneel in adoration	Psalm 103:1,2
Zamar	to make music (by touching the strings of an instrument)	Psalm 71:22
Shabach	to shout, to command in triumph, to speak well of	Psalm 147:12
Tehillah	to sing a hymn or a halal of praise	Psalm 40:3
Towdah	to praise by giving or extending the hand with an offering	Psalm 50:23

Without going into an exhaustive word study, it is clear that what is conveyed by the word "praise" seemingly contradicts the religious protocol of a huge majority of our churches. Arguments against praise of this sort, to my mind, reflect the tendency of many churches to have a spectator, rather than a participatory, mindset. Furthermore, to refer to that kind of praise as "Old Testament theology" mocks the unchangeable nature of God. Praise is ordained by God to go on for all of eternity.

The *Westminster Shorter Catechism* has served as a concise presentation of biblical truth in many churches including the church I grew up in as a young person. The definitions for praise given above support the catechism's answer to the question regarding the chief end of all mankind:

Man's chief end is to glorify God and enjoy him forever.[44]

To glorify means to honor, to praise or to exalt. Not only is praise ordained by God, it is part of God's plan and purpose regarding the whole of creation, mankind in particular, because only we of all creation have the power to choose to praise him. This fact makes the praise of humanity all that more pleasurable to God's ears.

How and why is praising God a part of who we are? As God was creating the universe, in which we live, one of the things that I believe amazed the host of heaven as they watched was the way in which all created things reflected the glory of God. Even as God's handiwork proclaimed his glory, the singing and the praise of host of heaven must have risen to a new level and an even greater volume. The prophet Isaiah was privileged to hear this song when he received his commission from The LORD of Hosts.

Holy, holy, holy is the LORD of Hosts, the whole earth is full of his glory. (Isa. 6:3)

The apostle John was also given a future glimpse of the host of heaven, joined by every creature in heaven and on earth, singing:

To him who sits on the throne and to the Lamb be praise and honor and glory and power forever and ever! (Rev. 5:13)

The most pleasurable and joyful work that the host of heaven engages in is that of praising God and his son Jesus Christ. We, created by his hand and for his glory, are called to mimic that work. Not only must we live in a physical atmosphere of oxygen and nitrogen and other elements, we must also reside in an atmosphere of praise in order to be spiritually healthy.

Although the admonition by God to join with all creation in praising him should be sufficient to prompt us into action, it often isn't. God in his mercy has revealed his glory and goodness so that we might see it, hear it, feel it and comprehend it. A responsibility lays on us to acquaint ourselves with God's goodness and to diligently observe his gentleness, grace, and love so that we are motivated to personally and individually engage in the practice of praise.

> When we see his works, when we hear his Word, when we taste his grace, when we mark his providence, when we think upon his name, our spirits [should] bow in lowliest reverence before him and magnify him as the glorious LORD.[45]

I can say with David that God has rescued me from the grave, he has restored me, he has provided for me. I have seen his goodness in the land of the living. God has been so abundantly good to me that my song will be never ending!

If praise comes to your lips with difficulty, perhaps you have not been diligent to observe God's goodness in your life. Have your spiritual ears begun to suffer hearing loss because of the debilitating decibel level of the clamor of this world? Scientists say that humans can only hear roughly one fifth of the audible sound spectrum. Perhaps our attention to what can be heard "in the spirit" has likewise suffered or been neglected.

Let me ask you another question. Do you perceive the context in which the volume of my song of praise grew louder? It was not when everything was just fine. It was in the midst of the storm! My hymn of praise grew out of deep need that caused me to seek God's presence. With hindsight and with new wisdom and experience I can see purpose in the trials I had to endure.

If we are willing to see, we will not lack for opportunities of beholding his goodness.[46]

Why praise? We should praise because the noise of this world desires to make humankind deaf to the song of all creation. We should praise because our hope and our desire, Jesus, is one day closer to catching us up into the glory of his presence. We should praise because "morning by morning" we see more and more of God's mercies renewed in our lives and in the lives of other believers. We should praise if God is greater in our lives today than he was yesterday. We should join others in praise, because our unified praise becomes a symphony of song.

To melody is added harmony and rhythm. The counterpoint of other melodies blends together with the unique timbre of many voices (some mined out of silver or brass, some crafted of ivory or fine woods, some molded out of composite materials) and rises up just like the perfumed smoke rose from the altar of incense as a fragrant offering unto God. Praise eloquently touches and speaks to the flesh and to the hearts, souls, and spirits of all humanity in a language that transcends words alone. It heralds the renown of our God and the wonder of his goodness to all who will hear.

Our praise is a witness and a personal testimony that identifies us as children of God. Praise is a garment that replaces the dingy, drab, and threadbare rags of despair. Praise ushers us into the throne room of his presence, into nearness and a place of intimate communion. Praising God with psalms and hymns and spiritual songs prepares us for service and is a sacred offering that is pleasing to God. We should praise, because when we do so, the Spirit of God anoints us with the oil of gladness, replacing sorrow and heaviness. Praise causes our faces to shine with God's glory and leads us to joyfully draw water from the wells of God's salvation.

When we become full of the joy of the LORD, we become strong. Praise silences the enemy and brings him to a standstill as one bound and fettered. Along with the Word of God, it is an offensive weapon that is so offensive to God's enemies that it causes them to flee. Praising God allows the host of heaven to execute vengeance upon those that hate his name.

Not only should we praise, we should praise continually—at work, at play, behind the wheel, at home, and in every place that daily living takes us. Continual praise will lift us above the shriek of howling winds. It will prevent the enslaving chains of discouragement from weighing us down and will liberate us from the heaviness of circumstances.

Let me encourage you to renew your song, your boasting, and your praise with these words from Charles Haddon Spurgeon:

> Just as the leader of an orchestra taps his baton to call all to attention and then to begin playing, so I now arouse and stir you to offer the sacrifice of praise to God.[47]

THE PRACTICE OF THANKSGIVING

Although the act of thanksgiving is one of the definitions of praise, God's Word talks so much about it that we ought to consider "Giving Thanks" as the inseparable twin of "Praise". Neither can go anywhere without the other. Just as praise draws us into God's presence, so thanksgiving lifts us nearer to him and is itself also a fragrant offering. If we look at praise as having a boisterous "rah, rah" character or a reverent adoration quality to it, giving thanks perhaps reflects more of a personal accounting for the favor that has been shown to us. Spurgeon's words eloquently communicate something of this partnership.

> A soul saturated with divine gratitude will continue to give forth the sacred aroma of praise, which will permeate the atmosphere of every place and make itself known to all who have a spiritual nostril to discern sweetness.[48]

God's Word gives us at least four reasons why we should zealously practice giving thanks.

1. GOD IS DESERVING

The following is an abbreviated list of how God is good to us individually, to our families, and to the body of Christ of which we are members.

Give thanks to the LORD for:

1. His love endures forever (Psalm 118:1)
2. His goodness and his everlasting mercy (Psalm 106:1)
3. The gift of Jesus Christ (2 Corinthians 9:15)
4. Christ's power and reign (Revelation 11:17)
5. The effectual (extremely efficient) working of his Word (1 Thessalonians 2:13)
6. Deliverance from sin (Romans 7:23-25)

7. Jesus' victory over death and the grave (1 Corinthians 5:57)
8. Wisdom and might (Daniel 2:23)
9. The triumph of the gospel (2 Corinthians 2:14)
10. Faith exhibited (Romans 1:8)
11. Love shown to us by others (2 Thessalonians 1:3)
12. The grace of God shown to us (1 Corinthians 1:4)
13. The nearness of God's presence (Psalm 75:1)
14. Divine provision (Matthew 6:25-34)

This list could go on and on. Several times in my life I have been challenged to make a list of things for which I am thankful (and which I carry around in my personal bible). After some thought, I realized I should begin by being thankful for a heritage of gracious, God-fearing love from my grandparents, who first touched my parents and then generously poured into my own life. I am also thankful for the loving support and Christian upbringing I received from my parents.

I am thankful for my own salvation, for my beautiful, loving wife and my wonderfully gifted children. For the talents God has blessed me with, for loving, faithful friends, for five loving church families over the last 30 years, and for their tremendously gifted and insightful pastoral staffs. And—all the way to the present—for the miraculous intervention of God in preserving my life. I believe that for every negatively perceived experience in our lives, we probably have at least dozens, if not hundreds, of things to be thankful for. We must petition God to open our eyes to the endless host of reasons we have to be thankful.

2. THANKSGIVING IS A KEY THAT OPENS DOORS AND GATES

Despite all the disasters David experienced in his life (older brothers who scorned him, issues with immorality and adultery, complicity in a murder, conflicts with his sons and daughters, enemies who constantly sought to kill him, betrayal, mocking criticism, and depression), he knew that giving thanks was a key that allowed him into God's presence. In one of the most upbeat, encouraging passages of scripture in the Bible, David writes:

> *Shout for joy to the LORD, all the earth. Serve the LORD with gladness; come before him with joyful songs. Know that the LORD is God. It is he who made us, and we are his; we are his people, the sheep of his pasture. Enter his gates with thanksgiving and his courts with praise; give thanks to him and praise his name. For the LORD is good and his love endures forever; his faithfulness to all generations. (Ps. 100)*

He knew something that Paul would write about hundreds of years later in his letter to the Philippians:

> *In everything, by prayer and petition, with thanksgiving, present your requests to God. (Phil. 4:6)*

Although we know God's hearing is not limited in any form or fashion, it's as if thanksgiving causes his attention to be drawn uniquely to that situation. True thanksgiving delights God's ears. I can imagine that as God smiles with joy when he hears our thanks, the host of heaven is breaking into shouts and cheers and joyful songs, thus amplifying our individual and congregational offerings of thanksgiving.

I'm sure many of us have spoken "thank-you's" to many people and in many ways. Occasionally a heartfelt thank you comes our way with a tear and a hug or a loving gesture of real appreciation. It moves our hearts and creates an unbreakable bond with the one we share that moment with.

After my liver transplant, our church family responded in numerous ways—with prayer, finances, meals, transportation, and by sitting with me so my wife could return to work. There exists a bond now that I can't ever imagine being broken except by moving to another geographic area.

Imagine how God's heart is touched by deep, sincere thanksgiving from those who recognize and acknowledge him and choose to bless him with such sacrifices! Now imagine his response to your loving, openhanded offering, and remember that God's love endures forever and is infinitely compassionate and resourceful.

3. THANKSGIVING SILENCES THE ENEMY

Psalm 107 begins with an admonition to give thanks to God for his goodness. Eugene Peterson's paraphrase of the next line is especially meaningful:

His love never runs out! (v. 1b The Message)

This whole psalm is about making thank-filled statements and confessions regarding God's love, his provision, his healing touch, his timely deliverance, and his ability to still the storm. Four times the psalmist returns to this chorus or refrain,

> *Let them give thanks to the LORD for his unfailing love and his wonderful deeds for men. (Ps. 107:8, 15, 21, 31)*

Near the end of the psalm, the writer reveals a powerful spiritual principle. The person who is wise will diligently observe the goodness of God and respond with thanks. But the wicked—the wicked will have no recourse but to shut their mouths!

Verbal offerings of thanksgiving combined with joyful praise are not only a powerful testimony, but they also become an invincible two-edged sword that inflicts vengeance and punishment on the enemies of God and his people. Not only will our enemy's mouth be silenced, but our thanksgiving and praise will

> *bind their kings with fetters and their nobles with shackles of iron,... [and] carry out the sentence written against them. (Ps 149:8. 9)*

This spiritual principle reminds us that it is not our responsibility to avenge ourselves or rail against our enemies. In times of conflict, disappointment, betrayal, or unexplainable illness we are to look at what God has done and be thankful for what he has done. We are to remember that his love, indeed, never runs out and that he is always faithful to those that are his, because he cannot be unfaithful to himself.

To return again to the analogy of a musical conductor, we must be tuned up and ready to follow his downbeat.

After the third click (of the baton) is heard, his hands are raised to signify an inhaling. When the conductor's hands come down, it will be the prompting to exhale a blast of the wind of the Spirit and jolt the enemies of God into submission to His [God's] authority in us.[49]

4. THANKSGIVING IS MEANT TO BE A SIGN

Not only do we have the personal benefits that a thankful heart offers to us, thanksgiving is also meant to be a testimony to all the world of the power, the love, and the grace of our God. Psalm 105 begins with these words,

Give thanks to the LORD, call on his name; make known among the nation what he has done. (v. 1)

These words are also quoted in David's psalm of thanks in 1 Chronicles 16, part of the incredible celebration ordained by David on the day that the ark was returned to Jerusalem. The giving of thanks, along with songs of praise, instrumental music, and the offering of petitions was to be an ongoing, daily ministry meant to honor the presence of God associated with the Ark of the Covenant. The musical, celebratory character of the offering of thanksgiving served to remind all of Israel and all who visited Jerusalem of the renown of their God.

New Testament admonitions to always give thanks (Eph. 5:20), to give thanks in all that we do (Col.3:17), and to give thanks in all situations (1 Thes. 5:18), are not commands to be thankful for flat tires or sickness or disaster. Rather, they are meant to be reminders to ourselves that God is always in control despite the wind and the waves. They are also reminders to the world that our lives are not ruled by what happens to us—good or bad.

Especially during times of difficulty, the world needs to know that our lives are firmly established on God and upon his Word. To those who do not know or have a relationship with God, our public offerings of thanksgiving are meant to show that our faith and trust are in God, and our advocate is our Father, Jesus Christ.

THE DISCIPLINE OF WAITING ON GOD

During the forty days after Jesus' death, he appeared to the disciples numerous times. Immediately after his resurrection, he instructed them to go to Galilee, where they stayed for about a month. Then he led them to Bethany, perhaps to the home of his friends Mary, Martha, and Lazarus. Just before his ascension into heaven, he gave the disciples some very specific instructions about returning to Jerusalem. They were to wait there until they received the promised gift of the Holy Spirit.

For ten long days they waited. I'm sure that during that time emotions ran both high and low. Who was to be in charge— Peter, James, John, one of Jesus' brothers, perhaps Mary, his mother? What should they do while they waited? Their dilemma is one that scripture tells us over and over again to embrace—to wait upon God.

Wait for the LORD; be strong and take heart and wait for the LORD. (Ps. 27:14)

Waiting is incredibly difficult for many, probably because we regard it as a hindrance or a roadblock. In our eyes, waiting prevents us from accomplishing what we have purposed to do. Of course, not only is that response a selfish one, it also reveals a critical lack of self-discipline and patience.

Unfortunately, the absence of these qualities often results in irritation and criticalness. An inability to wait often lays out a welcome mat inviting our enemy Satan and his spiritual forces to torment us even further by creating maddening delays that extend our wait even longer.

The act of waiting in the spiritual realm is not easy. We often believe we know what God ought to do in a given situation. Any delay results in an unsettling doubt regarding God's awareness of our circumstances rather than a reexamination of our self-centered desires...or a renewed acquiescence to God's will, his timing, and sovereignty. For most, the real agony of waiting lies in the fact that we feel we must be doing something. We seldom come to a place of complete stillness until we have exhausted our own efforts to solve our problems.

The record of scripture tells us that frequently God's plan requires his servants to wait on him. Look closely at the following list of biblical figures:

Table 4

Noah	Scripture doesn't tell us the time period between God's declaration that he was going to destroy the earth by flood and when it actually began to rain, but Noah had time to build a boat 450' long, by 75' wide, by 45' tall—all by hand, beginning when he was at least 500 years old! Then he spent 150 days waiting for the rain to stop and the water to recede.

Abraham	Abraham, then known as Abram, was 75 when he was first told that he would be the father of a great nation. When he was 99, the LORD appeared to him and told him that in about a year he and Sarah would have a son.
Joseph	Joseph was 17 when he had a dream in which his brothers bowed down to him. Shortly thereafter his brothers sold him into slavery. He was 30 by the time the Pharaoh of Egypt rescued him from jail and placed him in a position of authority. It was at least another 7 years before his brothers journeyed to Egypt to buy grain, and his dream came true.
David	David was anointed King over Israel when he was very young, probably a teenager. When he volunteered to fight Goliath, King Saul remarked that he was "only a boy." (I Sam. 17:33). He didn't become King over Israel until he was 30.
Jeremiah	Jeremiah's prophetic ministry regarding the fall of Judah spanned five decades.
The Disciples	Luke tells us in the book of Acts that Jesus appeared to the disciples over a period of 40 days after his resurrection. They followed his directions to return to Jerusalem (a short day's walk, Acts 1:12), where they waited for ten days in the upper room.

An examination of the word translated "wait" draws attention to the fact that what the writers of scripture (led by the

Holy Spirit) fully intended to convey is often missed. The prophet Isaiah writes,

> *Blessed are all who wait for him! (God) (Isa. 30:18)*

And again,

> *I will wait for the LORD. (Isa. 8:17)*

In both verses, the idea of adhering or sticking to God in times of adversity is commanded. His word says he longs to be gracious to us and show us his favor. Despite what feels like delays or even God hiding his face from us, there is great benefit to faithfully waiting on God. He cannot be unfaithful to himself, and we are part of him as his children. Those who desperately cling to God will be rewarded.

Isaiah again writes of the benefits of waiting in one of the most well known passages of the Old Testament,

> *But those who wait upon the LORD will renew their strength. They will soar on wings like eagles; they will run and not grow weary, they will walk and not be faint. (Isa. 40:31)*

Here and also in Psalm 37,

> *Those who wait on the LORD will inherit the land. (V: 9)*

Waiting implies being inextricably bound together like strands of a rope. One of the benefits of being "tied" together with God is an inner, spiritual strength that far exceeds the benefits of physical strength or mental and emotional resolve. Even the strongest among us grows tired. But our God never grows weary! When our lives are wound together with God and his principles, his strength will sustain us and renew us. The reference to inheriting the land refers to becoming the beneficiaries of God's promises.

David gives us another perspective on waiting in Psalm 68:

O my Strength, I watch [wait] for you. (Ps. 59:9)

Found here is the principle of narrowly focusing on God, on watching attentively as if standing guard. A vivid picture of this principle is also found in another Psalm:

I wait for the LORD, my soul waits, and in his word I put my hope. My soul waits for the LORD more than watchmen wait for the morning... (Ps. 130:5, 6)

Standing guard gives new meaning to the concept of waiting. That kind of focus requires mental, physical, and spiritual alertness rather than the vacant inattentiveness often associated with waiting. An alert sentry will not be caught off guard, will discern the signs of an approaching assault, and will be prepared to not only defend but repel the enemy's attack.

Another obvious meaning of the word "wait" means to attend to someone like a servant. Paul writes to the Corinthian church about this subject:

Don't you know that...those who serve [wait] at the altar share in what is offered at the altar? (1 Cor. 9:13)

Although in this passage he was specifically addressing their reluctance to support him and Barnabas financially, a bigger principle is illustrated. When we wait on God like a faithful server/servant, we will benefit from what he has to offer from his vast and infinite storehouse of grace. If we serve to receive material blessings, we are choosing poorly. We should serve/wait on God in order to build up the kingdom, and then the scripture says all the other less important things (that moths can eat and rust can ruin) will be given to us as well. (Matt. 6:33)

God's grace was poured out to my wife and family and me in overflowing measure during the most critical days of my illness. Gestures of love, offerings of prayer, and financial blessings came our way in overwhelming abundance. In the midst of this humbling generosity, one precious brother remarked, when we objected that he had already done so much, "Don't rob me of a blessing by taking away my opportunity to serve you!"

Our eyes were opened to see beyond what we were receiving to how God was moving in the midst of our church family and in the community where we lived. Even many months after the transplant, I met people who told me I was still on their personal or their church's prayer list. I will never cease to tell of God's amazing generosity, of his love and grace demonstrated in my life!

Another aspect associated with waiting involves having a sense of expectation. Paul writes in Romans that even though we have not yet received what we are waiting for, we are to wait eagerly and patiently. (Rom. 8:23, 25) The difficulty associated with waiting often results from a sense of doubt that what we hope and wait for may never happen.

Paul even writes that the earth itself, inanimate creation, is frustrated but waiting eagerly for the new creation to be revealed. If inanimate creation can wait, then so must we (who are called sons of God) discipline ourselves to look forward with eager anticipation to what will be revealed. God's word makes it clear that not one of his words will fail to accomplish its purpose, and therefore God cannot fail to rescue from sin and death those who believe on his name. That doesn't mean that no one will succumb to sickness and die; it means that death cannot separate us from God. Paul goes on to write further about the subject of our hope:

> *Who shall separate us from the love of Christ? Shall trouble or hardship or persecution or famine or nakedness or danger or sword? (Rom. 8:35)*

In a paraphrase of the balance of Chapter 8 of Romans, Paul indicates that absolutely nothing, not death or the fear of death, nothing that we are going through at this moment or anything that might happen in the future, no demonic torment, no crushing blow, or tidal wave of suffering can prevent us from receiving the benefits of God's love that are ours through Jesus Christ.

At this point, let me address the reality of pain, be it physical, emotional, or mental. Pain is real. Suffering is exhausting. Heartache is devastating. Despair is crushing. As I was struggling with the discomfort and pain associated with liver disease, it was difficult for me to think clearly and process all the truth that should have come forward from my spiritual heritage. It is at this point that the body of Christ, the Church, the fellowship of believers must step in and minister very real comfort to those who are struggling. Too often, people bring trite offerings of condolence that only reveal their desire to keep a safe distance between them and the one suffering. Alternatively, they sternly quote Bible verses that convey that really true faith would have kept the sufferer out of their situation. Rather, we should offer gentle touches of compassion, peace-producing gestures of comfort, generous gifts of time, mercy-filled moments of understanding, and faithful acts that demonstrate the oneness of companionship.

Job proclaimed,

> *I will wait for my renewal to come. (Job 14:14)*

We must, as believers, learn a lesson from Job's friends. Rather than waiting and tarrying in compassion with him, they pressed him with their criticism and condemnation. Rather than making his suffering easier, they increased his suffering and made his wait more difficult. Let us be ambassadors of "good will," compassionately and tenderly aroused to mercy.

There is one last understanding regarding waiting that must be understood. David writes in Psalm 62,

> *My soul finds rest in God alone... wait my soul, on God alone. My hope comes from him. (v. 1, 5)*

Sometimes waiting requires that we settle down and quiet our soul. God's direction to me required that I be silent and still. It was in that moment that I was able to enter into, not only a physical place of rest, but a mental and emotional oasis of peace as well. In all our "doing" we forget that it is in God that rest and salvation are found. Remember the verse from Isaiah quoted earlier,

> *In returning and rest shall ye be saved, in quietness and in confidence shall be your strength... (Isa. 30:15)*

What gift is available to every believer going through trials and times of difficulty? What should be the first thing we seek? The restful, peaceful, beautiful, renewing, and sweet smelling presence of God. In the midst of the wind and the waves, Jesus our advocate speaks these words,

Peace! Be still! (Mark 4:39)

And to us he offers this promise,

Come to me, all you who are weary and burdened, and I will give you rest. (Matt. 11:28)

In conclusion, let us reexamine the breadth of meaning conveyed in scripture's admonition to wait on God. Even though the act of waiting is difficult, we must discipline ourselves to yield to God's timing and bow our reason to his revelation, acknowledging that in the present we do not have to know his will. We must wait on God's solution, putting a stop to the frustration of trying to solve our problems in our own strength. Rather than spinning our wheels, we must cling desperately to him. We must inextricably wrap ourselves and all of what we do in God and in his word. Rather than shutting our eyes, we must refocus our attention on his promises. We must draw near to him and to the wells of salvation so that we may receive what he offers. We must be confident that what he offers is ours and that absolutely nothing can separate the giver and the gift from us. And lastly, we must learn to do all this while inhabiting the fragrant stillness of his presence.

When we have taken first one step and then another and another, following the path that God has cleared, graded, and prepared, when we have taken in the sights and smells, when our spiritual eyes and ears have reveled in the glory of what has been revealed along the way, we are ready to enter.

The best place to be is where God puts you.[50]

Fenelon

Rest is a place.[51]

Joyce Meyer

CHAPTER SIX

THE SECRET PLACE

He who dwells in the shelter [the secret place] of the Most High will rest in the shadow of the Almighty. (Ps. 91:1)

What and where is this secret place? It is obvious as you read further in Psalm 91 that the writer has been there. He speaks of shelter and security in the midst of peace-defying events and circumstances. While multitudes were perishing all around him, he experienced the comforting and trust-inspiring deliverance of Almighty God. He felt the hovering presence of God protecting him and the mighty armor of God's promises shielding him and quenching the searing heat of his enemy's attack. And most importantly, he learned to dwell in this shaded oasis of rest.

Many of us presume to have some sense of direction in our lives, yet when we find ourselves in the midst of sickness, tragedy, or crisis, it seems as if God is far away. We become discouraged or fearful, and we exaggerate our suffering. Mentally, emotionally, and spiritually we run wild, our imaginations so out of control that our situation seems worse than it actually is. Worry nags and eats away at us like a fire that refuses to be put out. Anxiety-producing thinking overwhelms us. Fear of events that might occur strain us to the breaking point. Our flailing about prevents us from sensing the tender touch of God's love. Our frantic gasping for air keeps us from inhaling the fragrance of God's presence. Our overactive minds scream, "Do something!"

It is at this moment and at this place of need that God offers us the opportunity to choose. Our response in our moments of crisis always involves choice. Author Elisabeth Elliot writes,

> We will accept either the solutions, answers, and escapes that the world offers or the radical alternative shown to the mind attuned to Christ's. The ways of the world exalt themselves against God. Choices will continually be necessary and let us not forget—possible.[52]

The *necessary* choice that is offered involves placing ourselves "under orders." Choices and actions based on our own human understanding are of little or no value. We must learn to choose to rest in God's hands. We must learn to live with God in the present moment, and we must accept by faith that he will give us everything that we need. The key to the secret place is found in surrender to God. Jesus said,

> *In this world you will have tribulation and trials and distress and frustration; but be of good cheer—take*

courage, be confident, certain, undaunted—for I have overcome the world. I have deprived it of [the] power to harm [you]. (John 16:33 Amplified)

The truth that is so difficult to embrace in these words is that God's sovereignty will hold us fast regardless of our circumstances. The fact of suffering tempts us to believe that God is not to be trusted, that he does not love us, and that He cannot manage the events of our lives as well as we can, especially when our circumstances do not seem to change after we have rebuked demons, claimed scriptures, prayed, and been prayed over. We want to impose some limits on our suffering and on our weakness. Paul records God's answer to his repeated prayers that he be delivered from his "thorn in the flesh,"

My grace is sufficient for you, for my power is made perfect in weakness. (2 Cor. 12:9)

Douglas Webster has written,

> The thorn is not there to make us feel sorry for ourselves and seek the sympathy of other believers. The thorn is not an obstacle to God's will, but a catalyst for doing God's will. A thorn…prepares us for ministry.[53]

Through this thorn Paul was constantly reminded to turn his face toward his Savior and LORD. God received greater glory because of Paul's thorn. Let us be reminded that Paul's thorn and those that we bear are but a single thorn compared to the crown of thorns that Jesus wore upon his brow. Early in Jesus' ministry, he chose to embrace the will of his Father.

> *My food [meat], said Jesus, is to do the will of him who sent me and to finish his work. (John 4:34)*

In Gethsemane, he proved the strength of his will by proclaiming,

> *Not my will, but yours be done. (Luke 22:42)*

And again, on the way to Golgotha, it was the will of Jesus that triumphed when pain and exhaustion caused him to stumble and fall under the weight of the cross.

Many times our crosses cannot be carried or dragged. It is when we fall down beneath their weight that we face the choice to surrender to our feelings, to the wisdom of the world, to our worries and our anxieties, or to surrender to God's power, God's timing, and God's will.

Francois de Fenelon, the 17th century French author of *The Seeking Heart* wrote,

> Behind every annoying circumstance, learn to see God governing all things. The intrusions that God allows to come your way will no doubt upset your plans and oppose all that you want. But, they will chase you to God. Sit still before him and yield your will to him.[54]

Even if we are not masters of our emotions and our thoughts, we are by the grace of God masters of our consent. Even though our minds and our hearts prefer arguments to obedience and solutions to truth, obedience and surrender to God is always possible.

Again, I quote author Elisabeth Elliot,

> God's Word does not explain everything necessary for our intellectual satisfaction, but it does explain everything necessary for obedience and thus God's satisfaction.[55]

We who choose to trust in God do not always need to know why, nor do we need to understand every mystery. Almost universally, the events of our lives require time and hindsight before we have any inkling of God's purpose. It is only by faith that this principle of surrender can be embraced. Despite our lack of knowledge about what will happen next, faith acknowledges that behind and beyond every circumstance, even when he appears hidden from us, God is in control. God wants us to trust him in every moment and in every situation. In the famous hymn, *Great is Thy Faithfulness,* we are reminded that, "Morning by morning, new mercies we see."[56] Hymn-writer Thomas Chisholm's inspiration for this line came from this promise,

> *His compassions never fail, they are new every morning. (Lam. 3:22, 23)*

We must choose to surrender to God and allow him to take care of his business. At this point, several questions arise. Doesn't scripture tell us that the gift of faith allows us to boldly approach God's throne and ask whatever we will in Jesus' name with the confidence that God hears and will answer? Are we not allowed to ask God to deliver us from our circumstances instead of quietly surrendering to them? Isn't the Christian supposed to be engaged in spiritual warfare and therefore warring against the forces of darkness in high places?

The answer to all of these questions is an emphatic, "Yes!" The principle of resting in the midst of trials is not a matter of giving in to sickness or allowing ourselves to be battered by our negative circumstances. By all means, claim God's promises, let the elders of the church anoint you with oil and pray over you. Daily put on the armor of God and war against the enemy using the sword of the Spirit.

The problem so many of us struggle with is God's timing. Obedience and faith the size of a mustard seed must not be thought fruitless when God does not respond immediately or if God responds with, "My grace is sufficient for you." In special moments, and according to God's grace, the miraculous does occur. Most often, when God sees fit to reach down from his throne and cause a bush to burn and yet not be consumed, when he closes the mouths of lions, when he feeds thousands with five loaves and a couple of fish fillets, he is ministering in a unique situation. When the miraculous occurs, our minds tend to think that God is giving us a formula, and if we will just follow the steps, then bang! our miracle will happen. That is not how God takes care of his business.

We must also be aware that God is sovereign. Think back to the story of Shadrach, Meshach, and Abednego. If you and I were in their shoes, I'm sure we would have been praying that God would prove us guiltless and that he would change the king's mind about the prescribed sentence for the crime of not worshipping the king's gods or the image of gold that he had set up. Instead, the three Hebrew young men surrendered to God. Read their reply in the face of their death sentence.

If we are thrown into the blazing furnace, the God we serve is able to save us from it, and he will rescue us from your hand, O king. But even if he does not, we want you to know that we will not serve your gods or worship the image of gold you have set up. (Dan. 3:17, 18)

They knew there was a possibility that God would require his grace to be sufficient. Scripture tells us several paragraphs later that they trusted God and were willing to let go of everything, even their lives. What was their unique reward? Their lives were spared in one of the most miraculous stories recorded in scripture.

As I meditate on the significance of their trust and the totality of their surrender, I am reminded once again of the fact that I frequently lack a complete understanding of "the big picture." When our prayers for deliverance are not prefaced with "your will be done," we are in effect putting a limit on God's hand. God often takes us down a different path than the one we choose for ourselves, often to spare us from future hardship and suffering. The prophet Isaiah writes,

The righteous perish, and no one ponders it in his heart; devout men are taken away, and no one understands that the righteous are taken away to be spared from evil. (Isa. 57:1)

Often, our trials are permitted by God to prepare us, like someone in training, for future ministry—to equip us with the patience to endure and the love that prompts us to reach out to mercifully rescue others. James writes,

The testing of your faith develops perseverance. (James 1:3b)

He goes on to say that when we have persevered, when with patience we have remained standing, grounded, and established like a house built upon a rock, we will receive the gift of life.

With this promise in mind, the question arises, "Is your hold on this life so strong that you fail to comprehend, much less appreciate the glory of the eternal life that awaits you?" Perhaps the uneasiness and despair that afflict so many of us are in reality bringing to light some remnant of self that has not been put to death. Perhaps, even as I did, many of us have an emphatic hold on our present lives or attach a disproportionate value to things that have little eternal value.

Perhaps we have no confidence that God is in control and feel that letting go will create an out-of-control downward spiral in our lives and those of our loved ones. Our unwillingness to rest in God's hands in the face of trials is a telltale sign that our old nature is emerging from its hiding place and striving to regain control! That "old man" desires to falsely color the circumstances of our lives, subsequently stirring up impatience, tension, depression, self-pity, hypersensitivity, and screams of protest.

Having said all this, we must come back around to the issue of right choosing. Can we afford to frantically stumble around in the midst of our storms, thus adding to our fear and multiplying our agony? Should we struggle to carry burdens that threaten to crush us, or should we approach the throne of God in open-handed surrender, laying down all our cares at his feet?

The necessary choice involves deciding to follow the stepping-stone path that God has prepared and laid out before us and that leads to the fragrance of his presence. Walking along this "narrow road" will enable us to wait patiently with assurance that God knows our circumstances and has made supernatural provision for us to endure.

The word of God must be hidden in our hearts. During my illness, I kept a journal of sorts. Although the person who bought it for me intended for me to keep a record of my thoughts, it became an outlet for me to write scripture verses, lyrics to several favorite hymns, and choruses and thought provoking quotes from a variety of Christian authors. The Psalms became especially meaningful, because David and other psalm composers and authors often voiced the very same thoughts that were going through my head. When I was able, I often wrote out whole psalms, substituting my name and first person pronouns where appropriate as if I were the author. (See Appendix for an example).

These words became my prayers, because they were filled with boasts and faith statements about who our God is and what he has done and will do, as well as cries for salvation and rescue. When extended times of prayer became impossible due to my inability to concentrate for any length of time, I would briefly repeat or read over and over one of the now-familiar verses recorded in my journal, allowing the words of God to speak peace and stillness to my mind and body.

Engaging in praise and thanksgiving will also prepare us to patiently wait on God. I often listened to the music of my favorite Christian artists, especially during the discomfort of recovery from the liver transplant. Just as David's music soothed the spirit of Saul, so music that is anointed by God will usher us into his comforting presence. Whenever possible, I shared how my faith was sustaining me through those trying times, giving praise to God. The outpouring of support from family and friends was a continual reminder to be thankful for God's faithful and compassionate love and the "new mercies" that were evidenced in my life day by day.

The goal of this book is to encourage each of us, myself included, to search for and take up residence in the secret place of God's fragrant presence. Fragrance was the sense that God used to remind me of the pleasure of resting in the security of his love, his provision, and his protection. We must learn to let go of this earthly life's times of struggle, uncertainty, and difficulty and sit quietly before him, sure of his sovereignty and sure of his presence. We must remember that giving in to worry will dull our awareness of the presence of God. We must remember that faith is not dependent on the condition of our minds, hearts, or bodies, nor is it limited by those conditions. We must be aware that our confessed weaknesses and failures do not make us displeasing to God, but rather allow his strength to be demonstrated in our lives.

We must also remember to bear in peace the feelings that overwhelm us even after we have determined to sit patiently before the LORD, knowing that in God's time they will go away. We must remember that sitting in God's presence will bring softness, humility, restraint, and a sense of calm, despite the pounding of waves of despair and the might of stormy winds. We must make time for not only sitting quietly before him but also hiding his word in our hearts and reaching out to him in prayer, remembering that he gives as only God can give.

Our lives must become fragrant with the perfume of communion. We must fill our mouths with thanksgiving-filled praise and joyful shouting that cannot be silenced. We must learn to peacefully, patiently, alertly and expectantly wait on God, inextricably binding our lives, our words, our thoughts and our hopes to him. We must remember that even God rested from his work and decreed the sacred observation of times of rest. We must realize that rest can only be found in God's presence, because it is in him that we find the source of all that is good, right and pleasant. If we would reside in the shelter of his peace, we must lose sight of our selfish desires and rest in a spirit of non-resistance before the presence of the one who is "Alpha and Omega."

Nicholas Herman, also known as Brother Lawrence, wrote this powerful, yet practical summation in his book *The Practice of the Presence of God*:

> The LORD does not really lay any great burden on us. He only wants us to recall him to mind as often as possible, to pour out our adoration on him, to pray for his grace, offer him our sorrows, return from time to time to him, and quietly, purely thank him for the benefit he pours out on us even in the midst of our troubles. The LORD asks us to let him be the one who consoles us, just as often as we can find it in us to come to him.[57]

As we come to the conclusion of my efforts to lead you to a place of resting in *The Fragrance of Paradise*, hear God's Holy Spirit whispering to your spirit: Peace, peace, to those far and near. From heaven's heights, I have established peace, and it is from my sanctuary that I grant peace. From the farthest corners of all of creation, from the north, the south, the east and the west I have called you by name—you are mine! Although I live in a high and lofty place, I will come down from my dwelling to guard you day and night and to live with you who are contrite and lowly in spirit. Do not rely on your own understanding, rather put your trust and hope in me. I will bless you, direct you, and make your ways straight if you recognize and acknowledge me and if you rely on the wisdom of my words. I will order the steps of those that love my words so that nothing can make them stumble. All my ways are peace!

    Turn to me all you who are weary, all who are burdened, all who can barely whisper a prayer, all whose pain has caused you to cry out, and I will come to you and give you rest! Everything that could steal your peace (your anguish, your burden of sorrows, and the weight of your transgressions) has been laid upon me. I have gone before you to shield you from the stroke of the sword and the lash of the tongue. No weapon forged against you will prevail! When you pass through the strong currents of deep waters, they will not sweep over you! When the fire of tribulation seeks to sear your heart, I will be your shade! When anxiety attacks on every side, my consolation will be your joy and your strength. When you cry, "My foot is slipping," my love swill support you! I will silence the song of the ruthless and the roar of the destroyer. Do not fear, for I am with you! Do not be dismayed, for I am your God! For my own sake I do this; how can I let myself be defamed? I will not yield my glory to another. Who is my equal? I never grow tired or weary!

    Only one thing is needed. Set aside the distractions that have stolen your heart and mind, and sit at my feet. Turn to me and be saved. Remember, the battle is not yours. Trust me steadfastly, and I will keep you in perfect peace, for I have ordained peace for you! Cast your cares on me, and I will sustain you. Hurry to the place of my shelter, far from the tempest and storm. In my gentleness you will find rest for your soul. With my hands I will bind up and heal your wounds. My touch is gentle— in my arms I will carry you close to my heart. I will take hold of your hand and give you breath and life. I will keep you; I will free you from the confinement of captivity and release you from darkness into light. My wings of protection are spread over you; rest in the comfort of my shade. I have engraved you on the palm of my hands. You are mine, the one in whom I will display my splendor. My banquet table is set before you despite the presence of your enemies, and my banner over you is love. With songs of deliverance I will protect you. At night my song is with you and with my love I will quiet you. I will harden you to difficulties and give you strength by my righteous right hand. Because I am with you, you will rest secure and not be shaken. I will never leave you nor forsake you.

Wait patiently for me. Lie down and rest in peace. Close your eyes. Be still and take a deep breath. Inhale the fragrance of my presence. Let your restlessness be replaced with the stillness of my peace.

APPENDIX

Psalm 121 (KJV)

I will lift up mine eyes unto the hills, from whence cometh my help.
My help cometh from the LORD, which made heaven and earth.
He will not suffer thy foot to be moved: he that keepeth thee will not slumber.
Behold, he that keepeth Israel shall neither slumber nor sleep.
The LORD is thy keeper: the LORD is thy shade upon thy right hand.
The sun shall not smite thee by day, nor the moon by night.
The LORD shall preserve thee from all evil: he shall preserve thy soul.
The LORD shall preserve thy going out and thy coming in from this time forth, and even for evermore.

By simply substituting "me" for "thee" the promises found in this psalm become very personal. As you read and pray these words God's truth will be hidden in your heart, mind, and spirit, becoming your "shield and buckler." I encourage you to find a favorite psalm (or any other scripture), write it out and make it "yours."

Psalm 121 (Paraphrase)

When in despair I find myself looking for help - I look up - remembering that rescue comes from the LORD, the creator and sustainer of all.

He holds me fast, indeed his gaze is always fixed on me and everyone of His loved ones.

Ever vigilant, never weary, God is my loving keeper, close at hand.

His cooling shadow shades me from the hot breath of the enemy, his warming presence comforts me when all is cold and dark.

The LORD of all things, the ruler over all is my guardian and protector.

My mind, my heart, my whole being is in His care.

As He has kept watch over my past, my now and my future are secure in His hands.

Endnotes

Chapter One:
[1]*Garden Design,* p. 13.
[2]*So Far*

Chapter Two:
[3]*The Garden of Eden,* p.54, 55.
[4]*Heaven,* p.55.
[5]*Garden Design,* p. 22.
[6]*The Garden of Eden,* p. 47.
[7]Ibid, p. 54, 55.

Chapter Three:
[8]*American Rose Annual,* p. 88.
[9]*Plants of the Bible,* p. 82.
[10]Ibid, p. 75.
[11]Ibid, p. 39, 40.
[12]Ibid, p. 75.
[13]Ibid, p. 157, 158.
[14]*The Book of Potpourri,* p. 6.
[15]*The American Rose,* p. 7.
[16]*Dictionary of the Hebrew Bible,* p. 74.
[17]*Dictionary of the Greek Testament,* p. 78.
[18]Ibid, p. 20.
[19]*Plants of the Bible,* p. 82.
[20]Ibid, p. 75.
[21]Ibid, p. 148.
[22]Ibid, p. 87.

[23] "Tell Me About Costus,"
[24] *Plants of the Bible*, 76.
[25] Ibid, p. 224, 225.
[26] Ibid, p. 223, 224.
[27] Ibid, p. 102.
[28] Ibid, p. 57.
[29] *The Feasts of the LORD: God's Prophetic Calendar*, p. 51.
[30] *Plants of the Bible*, p. 102.
[31] Ibid, p. 82.
[32] Ibid, p. 82.

Chapter Four:
[33] *The Companion to Roses*, p. 176.
[34] *The Man in the Mirror*, p. 83-85.
[35] Unknown
[36] *Our Utmost For His Highest*, July 18.
[37] *An Enemy Called Average*, p. 169.

Chapter Five:
[38] *Garden Paths*, p. 97.
[39] *The Scented Garden*, p. 15.
[40] *The American Heritage Dictionary*, p. 439.
[41] *The Bait of Satan*, p. 12.
[42] *The Practice of Praise*, p. 161.
[43] *Possessing the Gates of the Enemy*, p. 181, 182.
[44] *Westminster Shorter Catechism Illustrated*, p. 7.
[45] *The Practice of Praise*, p. 151.
[46] Ibid, p. 135.
[47] Ibid, p. 167.
[48] Ibid, p. 150.
[49] *Sound of Heaven*, p. 154.

Chapter Six:
[50] *The Seeking Heart*, p. 85.
[51] *Be Anxious For Nothing*, p. 82.
[52] *Discipline, The Glad Surrender*, p. 43.
[53] *The Discipline of Surrender*, p. 98
[54] *The Seeking Heart*, p. 14.
[55] *Discipline, The Glad Surrender*, p. 11.
[56] *Hymns of Glorious Praise*, p. 12.
[57] *Practicing His Presence*, p. 71.

Bibliography

"Attar of Roses," *American Rose,* September, 1994

Bevere, John, *The Bait of Satan,* Lake Mary, FL:Creation House, 1994

Brother Lawrence, and Laubach, Frank, *Practicing His Presence,* Sargent, GA:The Seedsowers, 1973

Chambers, Oswald, *My Utmost For His Highest,* Edited by James Reimann, Grand Rapids, MI:Discovery House Publishers, 1992.

Douglas, Frey, Johnson, Littlefield and Van Valkenburgh, *Garden Design:History, Principles, Elements, Practice,* New York, NY:Simon and Schuster, 1984

Duke, James A., Ph. D., *Herbs of the Bible, 200 Years of Plant Medicine,* Loveland, CO:Interweave Press, 1999.

Elliot, Elizabeth, *Discipline, The Glad Surrender,* Grand Rapids, MI:Fleming H. Revell, 2000

Fenelon, *The Seeking Heart,* Sargent, GA:The Seedsowers, 1992

Fisher, John, *The Companion to Roses,* Topsfield, MA:Salem House Publishers, 1986

Hayward, Gordon, *Garden Paths, Inspiring Designs and Practical Projects,* Charlotte, VT:Camden House Publishing, 1993

Hughes, Ray, *Sound of Heaven, Symphony of Earth,* Charlotte, NC:Morningstar Publications, 2000

Hymns of Glorious Praise, Springfield, MO:Gospel Publishing House, 1969

Jacobs, Cindy, *Possessing the Gates of the Enemy,* Grand Rapids, MI:Chosen Books, 1994

Mason, J. L., *An Enemy Called Average,* Tulsa, OK:Honor Books, 2003

Meyer, Joyce, *Be Anxious For Nothing,* Fenton, MO:Warner Books, 2002

Mitchell, Jonie, "Woodstock," *So Far,* Atlantic Recording Corp., CD, 1974

Miller, N. F., "Study of Rose Fragrance," *American Rose Annual,* Columbus, OH:American Rose Society, 1962

Moldenke, Harold N., and Alma, L., *Plants of the Bible,* Waltham, MA:Chronica Botanica Company, 1952

Molinos, Michael, *The Spiritual Guide,* Sargent, GA:The Seedsowers, 1982

Morley, Patrick, *The Man in the Mirror,* Grand Rapids, MI:Zondervan, 1997

Prest, John, *The Garden of Eden; The Botanic Garden and the Re-Creation of Paradise,* New Haven, CT:Yale University Press, 1981.

Spurgeon, C. H., *The Practice of Praise,* New Kensington, PA:Whitaker House, 1995

Strong, James, *The Exhaustive Concordance of the Bible,* Nashville, TN:Abingdon, 1890 (1978)

The American Heritage Dictionary, 2nd College Ed., New York, NY:Houghton Mifflin Company, 1991

The Good Scents Co., *Tell Me About Costus,* http://acupuncturetoday.com/herbcentral/costus.html (accessed July 2, 2005)

Verey, Rosemary, *The Scented Garden,* New York, NY:Random House, 1981

Webster, Douglas D., *The Discipline of Surrender,* Downers Grove, IL:Inter-Varsity Press, 2001

Whitecross, John, *The Shorter Catechism Illustrated,* London:Banner of Truth, 1968

LaVergne, TN USA
06 November 2009
163308LV00001B/5/P